HEAVENLY ERRORS

Heavenly Errors

MISCONCEPTIONS ABOUT
THE REAL NATURE OF
THE UNIVERSE

Neil F. Comins

Columbia University Press
New York

FIGURE SOURCES

I.1—Yerkes Observatory

1.1b, 1.3, 1.7, 1.8, 2.2, 3.2—*Discovering the Universe*, Neil F. Comins and William J. Kaufmann, III, © 2000 Neil F. Comins; reprinted by permission, W. H. Freeman & Co.

1.6—Yerkes Observatory photograph; diagram from *Discovering the Universe*, Neil F. Comins and William J. Kaufmann, III

1.2—*Universe*, William J. Kaufmann, III and Roger A. Freedman, © 1999; reprinted by permission, W. H. Freeman & Co.

1.4—Adapted from *Astronomy Today*, Eric Chaisson and Steve McMillan, © 1996 Prentice Hall

2.1—Prof. Joel Mintzes, private communication

3.3—William C. Keel, University of Alabama, Lowell Observatory

3.5, 6.3—NASA

5.1—Adapted from *Foundations of Astronomy*, Michael A. Seeds, © 1999 Wadsworth

Columbia University Press
Publishers Since 1893
New York   Chichester, West Sussex

Copyright © 2001 Neil F. Comins
All rights reserved

Library of Congress Cataloging-in-Publication Data

Comins, Neil F., 1951–
Heavenly errors : misconceptions about the real
nature of the universe / Neil F. Comins.
p. cm.
Includes bibliographical references and index.
ISBN 0–231–11644–6 (cloth : acid-free paper)
ISBN 0–231–11645–4 (pbk. : acid-free paper)
1. Astronomy. 2. Errors, Popular. I. Title.
QB44.2.C65 2001
520—dc21   00-050853

Casebound editions of
Columbia University Press books
are printed on permanent and
durable acid-free paper.

Printed in the United States of America
Designed by Audrey Smith
3 5 7 9 c 10 8 6 4
1 3 5 7 9 p 10 8 6 4 2

Dedicated to my students,
past, present, and future.

CONTENTS

PREFACE

Few things in nature conform to our expectations: objects with different weights fall at the same speed; time slows down the faster you move; the bulk of a tree's matter comes straight out of the air. Scientific understanding reveals that space, time, matter, and energy are by and large inconsistent with our common sense. How can this be?

Our brains evolved to help us survive, not to comprehend the cosmos. But as a by-product of that evolution, we have minds that want, or rather, need to understand how the natural world operates. While this desire has existed far back into prehistory, only in the past few centuries have we developed the mental discipline of using science to bridge the gap between what appears to be happening and what really is happening.

Over the past decade I have been exploring the differences between appearance and reality in nature, especially in my own field of astronomy. For example, in the brief glimpses I have of it, I see the Sun as yellow. If I didn't know better, I would conclude that it is giving off just yellow light or, perhaps, mostly yellow light. Neither of these conclusions is correct. It is giving off mostly turquoise light. There are literally thousands of commonly held beliefs about nature that are wrong for a variety of reasons. I have written this book to explore some of these differences between reality and perception.

It is important to know that the differences between how nature really works and our beliefs about how it works is an active, volatile field of study, carried out today by thousands of educators, psychologists, and others around the world. Every aspect of it is under intensive scrutiny: identifying the common incorrect beliefs we hold, identifying their origins, finding ways to replace them with correct knowledge, developing techniques for avoiding them in the future. The work has resulted in thousands of publications and diverse ideas put forward on each of these topics, many at odds with each other. I do not intend to present a complete survey of all this impressive work. Rather, I have tried to highlight the points that I believe are most illuminating and helpful to anyone interested in better understanding the differences between scientific reality and perception.

I am greatly indebted to the more than eight thousand students to whom I have had the privilege of teaching introductory astronomy. They and my own two children provided me with the initial insights into how different our beliefs—in the realm of science, at least—are from reality. Professor Joel Mintzes deserves special acknowledgment for his willingness to let me use concept maps he created that contain incorrect beliefs. It is rare in any field for an expert to acknowledge such beliefs, much less share them with others, even if constructively. I also want to thank my colleagues at various institutions around the world who have provided me with considerable insight into various aspects of this realm of learning. Thanks also to Neil DeGrasse Tyson, Alex Filippenko, Jeanne E. Bishop, and Joseph Nuth, who reviewed and made suggestions to improve the proposal and early drafts of this book, and to my friends at W. H. Freeman, Publishers, who kindly let me reproduce for this book figures from my textbook, *Discovering the Universe*, 5th edition, by Neil F. Comins and William J. Kaufmann III, and from *Universe*, 5th edition, by William J. Kaufmann III and Roger Freedman. Special thanks to my wife, Suzanne, and children, James and Joshua, for providing feedback about this book and for putting up with me hard at work on it, and also to Holly Hodder, Publisher for Science at Columbia University Press, for her support of my writing over many years.

HEAVENLY ERRORS

Abandon common sense, all ye who enter here!

Introduction

The Mayall Four-Meter Telescope, one of the premier optical tele-scopes in the world, dominates the landscape at the Kitt Peak National Observatory, southwest of Tucson, Arizona. This telescope towers eighteen stories above the mountain peak and on every clear night the dome slides open, technicians aim the telescope, and astronomers gather new information about the cosmos. To avoid getting the tele-scope wet in unexpected rain, astronomers must periodically check for clouds. In the 1980s, this was always done by walking around a cat-walk on the outside of the dome. I performed this ritual many times, and invariably, at three in the morning, it had the effect of altering my perception of reality.

Leaving the quiet efficiency of the control room with its myriad

computers, data storage devices, and instruments for guiding the telescope, you quickly reach the access door, a steel affair more like the
hatch on a ship than anything else. When the door squeaks open and
you step outside, you are hurled into blackness ten stories above the
ground, as you stand on a fragile-looking steel catwalk. Indeed, it and
the guardrail are all but invisible at night, giving the unsettling impression that you are suspended in space. Though I had overcome my fear
of heights by jumping out of airplanes in the military, my first stroll
around that catwalk in darkness was nevertheless profoundly disori-
enting.

The circuit, longer than two football fields, took about five minutes
to complete. The Milky Way blazed overhead, its stars barely twinkling
(astronomers hate twinkling because it blurs images). The stunning
beauty and the fact that I was standing above a mountaintop reaching
high into the sky made me feel more connected to the universe than
ever before. I was reminded of the experience years later when seeing
the response of Jodie Foster's character, Dr. Ellie Arroway, to the view
of the cosmos she had on the other world in the feature film of Carl
Sagan's *Contact*.

As I walked around the dome I saw cloud-to-cloud lightning
flashes off to the north. This is one of nature's wonderful, free light
shows. Those clouds, moving eastward, presented no threat to my
night's observing. Down below, a car drove slowly up the winding
mountain road. It had on only its parking lights, so as not to interrupt
the work of the telescopes. I could see it through the grating at my feet,
which added to my sense of hovering above the ground.

Walking in nature's cathedral, with its wonderful spray of light, I
was intensely aware of how much still remains to be learned about our
cosmic environment. Every planet, moon, asteroid, comet, star, interstellar cloud, galaxy, quasar, and black hole, to name a few astronomical objects, has something to teach us, and I was thrilled to be involved
in these discoveries. But that involvement came at a price. Like all
astronomers, I have certain beliefs about how various parts of the cosmos work. Much more often than not, new observations require my
colleagues and me to modify or replace cherished beliefs. It is a price
most of us are willing to pay to better understand reality.

Later that week I was back at the University of Maine, in front of 250 students eager to learn more about the universe. Every semester for ten years I had taught such a group the latest information about astronomical discoveries and insights. My teaching was based on the assumption that if I said things clearly, students would digest that information, replace incorrect ideas they might have, and make the correct knowledge part of their understanding of how nature works. This intellectual house of cards collapsed that day.

It was the beginning of the second lecture of the semester. The day was sultry. I was acutely aware that the ventilation was entirely inadequate, that heads were nodding, and that the shirt I wore was getting soaked: "I'm melting. I'm melting." The class laughed. I had their attention again, but its shift away from their physical discomfort to a zanier-than-usual professor resonated with something in my subconscious. Like the change in perception I had experienced at Kitt Peak stepping out of the four-meter's dome, my view of the class was different. I could see myself from the perspective of a student sitting in my audience, listening to me.

I was showing them the constellations and the motion of the Earth, and had reached the point of explaining that there are actually thirteen constellations through which the Sun passes each year. "What is that guy talking about?" I thought, mentally watching myself from their vantage point. "I know there are twelve zodiac constellations. They're listed in the newspaper every day." I looked into their faces. So many of them, it seemed, were thinking that very thought.

I had been so naive about the learning process. The students weren't accepting what I was saying just because I said it. I had neglected to take into account the ideas and beliefs that these people had formed about astronomy and the rest of the natural world prior to my class. They were comparing what I said with what they already believed. Over the next few lectures, I carefully listened to myself from their perspectives. Dozens of concepts I presented sounded debatable because other explanations that many of the students might already have developed seemed more plausible. For example, I showed photographs of the Moon in different phases to prove that the same side of it always faces the Earth. These pictures show the same craters always in the same places throughout the cycle of phases (figure I.1). It seemed plausible to the "student" in me that for the same face of the Moon to stay oriented toward us, the Moon must not rotate or spin on its axis. Indeed, I remembered believing that when I was in college.

However, the Moon *does* rotate, at exactly the same rate that it orbits or revolves around the Earth. You can convince yourself of this

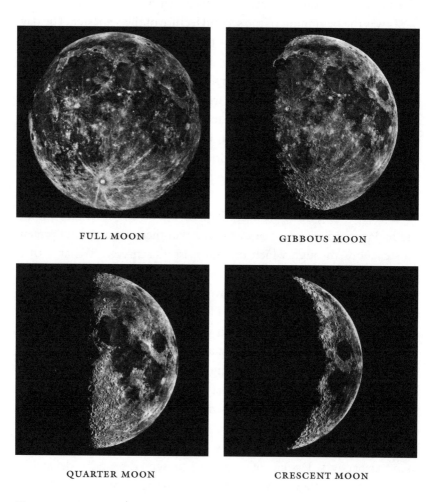

FULL MOON

GIBBOUS MOON

QUARTER MOON

CRESCENT MOON

FIGURE I.I Rotating Moon. You see the same features in all of these photographs because we always see the same side of the Moon, regardless of lunar phase.

by putting your arm straight out and bending your hand so that your palm faces you. Your palm represents the side of the Moon facing Earth. Move your arm horizontally around you keeping your palm (same face of the Moon) facing you. The only way you can do this is to have your palm changing direction (Moon rotating) as your arm moves around you. One revolution of your arm around your body (Moon around the Earth) must coincide exactly with one rotation of your palm. Otherwise, your palm (same face of the Moon) won't always face you (Earth). Reality and common sense are often at odds, but such demonstrations are a valuable tool in reconciling them.

Here are some questions about concepts that are often misunderstood. The answers are at the bottom of page 6.

Are seasons caused by the Earth's changing distance from the Sun?

Do comet tails trail behind the comets?

Do all the planets have solid surfaces?

Does the Sun shine by burning some gas?

Is the Sun a star?

Does the Sun move through space at millions of miles an hour?

Do stars rapidly change brightness (twinkle)?

Are black holes like giant vacuum cleaners in space, sucking up everything around them?

Are black holes black?

Are they holes?

Explanations of the answers to many of these and other questions asked below will be presented later.

Common incorrect beliefs are not limited only to things we have learned incorrectly or have misinterpreted. Our reasoning is often faulty. Consider this example: a year on Earth is the time it takes our planet to orbit the Sun once. Suppose that the Earth, still orbiting the Sun at its present distance, had only half its present mass. (An object's mass is the total number of particles it contains, without regard to the object's size.) Would the length of the year decrease, increase, or remain the same? As in the case of falling objects of different weights (a situation discussed in the preface), we might assume that an Earth of lesser mass would travel at a different speed. The answer is at the bottom of page 7.

In the course of living, we acquire or create explanations about things we see or hear in astronomy (and every other aspect of life, of course). Implicitly we accept these explanations as correct—otherwise we wouldn't believe them. This process begins when we are infants and continues until we die. But are the ideas we come to believe always right? Clearly not. Otherwise we wouldn't have so much difficulty learning things in school that we think we already know. The first step in correcting our knowledge is facing the tough reality that some of the information in which we firmly believe is wrong.

This is, of course, incredibly hard, so we all go to extravagant ends to change new facts to fit our beliefs, rather than vice versa. For example, suppose you believe that the planets have circular orbits around the Sun (which they don't). Now suppose I, in my expert role as an astronomer, tell you that Pluto, usually the farthest planet from the Sun, is sometimes closer to the Sun than Neptune. Do you throw out your idea about circular orbits? Probably not. Instead, your mind goes to work trying to reconcile your belief with the new fact about Pluto and Neptune provided to you by an expert. One common resolution to this potential contradiction is that the (allegedly) circular orbits of Neptune and Pluto don't have the same center, like two of the rings on the Olympics logo. The Sun is at the center of only one of the orbits, you might argue, keeping your belief in circular orbits intact.

Clearly, other people don't have the same understanding of astronomy that astronomers do. What do they think? And where did they get these erroneous ideas? While teaching that astronomy class, I thought it would be interesting to find out. At that moment, I began a decade-long journey into the amazingly convoluted worlds created by our minds. These are realms not only of sight and sound but also of minds reasoning by their own rules, creating wondrous mental land-

---

No; No; No; No; Yes; Yes; No; No; No; No

scapes with few signposts to guide others. Exploring what people think about astronomy was part voyage of enlightenment, part nightmare. Sometimes it was both.

Consider one incident. To better understand what people think about astronomy, I held meetings with small groups of students, typically half a dozen, to discuss their beliefs prior to taking my class and how their understanding had changed as a result of it. Black holes are an exotic and intriguing topic that many people encounter during childhood or adolescence. In the midst of a discussion on black holes, one student casually mentioned that as a child he had been told by his priest that if he misbehaved he would be punished by being sent into one. I felt my connection to reality slipping away: theology had adapted a scientific concept to its own use. I suppose I shouldn't have been so surprised, since innumerable science fiction stories use black holes in their plot lines. Adapting science to non-scientific situations is an established role of science fiction. But somehow this was different.

Driving home that afternoon, I felt a chill run through me as I realized the implications of the threat that had been made to that child, and, I learned later, to others. Imagine that you are an eight-year-old who has recently learned (correctly) that black holes are objects that things cannot leave once they have entered. Now consider your response to the threat of being sent into such an object. First is the terrifying prospect of being trapped forever and never seeing your parents and friends again. This is bad enough, but I submit that you would also make a connection between this fear and science, from which knowledge of black holes originated. And so science becomes linked with evil and is thereafter suspect for you.

I find it fascinating that we live in a natural world that rarely works as we believe it does, yet most of the time we function very well. I've written this book to explore the issues surrounding this discrepancy, and perhaps to lighten the load of numerous incorrect beliefs we carry.

The Earth's year would remain unchanged because the mass of a body does not affect its orbital motion around the Sun.

I

Fun in the Sun

SOME MISCONCEPTIONS
CLOSE TO HOME

We are exposed to more information about the solar system (the Sun and everything that orbits it, namely the planets, moons, asteroids, meteoroids, and comets) than about more distant space objects. Most of us observe solar system objects like the Moon, shooting stars, comets, and planets in the night sky, and astronomical observatories and spacecraft provide tantalizing news snippets about these and other phenomena in our cosmic neighborhood. This torrent of information carries many opportunities to develop misconceptions about astronomy. In fact, over two thirds of astronomical misconceptions pertain to objects in our solar system, with the rest centered around more exotic, distant objects in space, like black holes, galaxies, and quasars.

On one hand, it seems obvious that we manufacture so many incorrect beliefs about the solar system—after all, it is comprised of the most accessible elements of astronomy. On the other hand, it seems odd that the solar system is a source of so many wrong ideas—after all, astronomers understand so much more about its objects than about objects more distant and strange. As we will see in the next two chapters, the reasons for our misconceptions are nearly as varied as the features in the night sky.

Loosen Your Asteroid Belt

The late 1960s and early 1970s were NASA's heyday. Successful missions to the Moon, Venus, and Mars laid the groundwork for Pioneer 10 to be launched on March 2, 1972, en route to Jupiter and beyond. To get to these distant bodies, the 570-pound, 10-foot-diameter spacecraft had to cross the gulf between Mars and Jupiter, a region containing the asteroid belt. What dangers did it face in the belt, and how did NASA equip it to cope with them?

Except for NASA scientists and a few other astronomers who studied asteroids, most scientists recognized the asteroid belt only as a relatively obscure region of the solar system. Astronomers did know that the asteroids are primarily rocky and metallic debris left over from the formation of the solar system 4.6 billion years ago. Some asteroids, like Ceres, are spherical and up to one quarter the diameter of our Moon. But the vast majority are just a few meters across.

The Belt According to George Lucas

Before answering the question of the dangers that might have befallen Pioneer 10, let's consider the image of the asteroid belt in the popular mind today. This image dates from 1980, when the film *The Empire Strikes Back* was released. In it, we were treated to the memorable scene of the *Millennium Falcon* dodging asteroids to evade the pursuing Empire spacecraft. The asteroid field shown in the movie allegedly existed far from our solar system. Nevertheless, virtually overnight, most of the people on Earth conceived of our own asteroid belt as strewn with asteroids so big and so close to each other that you could practically jump from one to another.

George Lucas's portrayal of closely packed asteroids served an important cinematic purpose, but it also created a belief in the minds of countless millions (even billions) of people that is absolutely wrong. Consider what would happen if the *Star Wars* scenario were correct: each of those closely packed asteroids would be gravitationally attracting all the asteroids around it. If they were as large as portrayed in the movie, within a few thousand years of forming the belt, the asteroids would have smashed each other into dust or smashed into each other to form a single large body (which is how the planets in our solar system actually did form), or they would have had many near misses. Such near misses often have the effect of speeding one body up while slowing the other down. So an asteroid belt that didn't pulverize itself or form one larger body would quickly evaporate because after the near misses, the higher-speed asteroids would fly away from the rest. Therefore, clumps of asteroids as shown in the movie should not be found in the real asteroid belt.

But this is just a scientific theory. There might be other effects not accounted for here, and George Lucas could be right after all. What is the observation-based reality of the asteroids?

Passage to Jupiter

Let's return to 1972 and Pioneer 10. Most astronomers who studied the asteroid belt back then were convinced by scientific arguments like the one above that the asteroid belt must be nearly empty. But they were human like the rest of us and prone to asking "what if?" questions, like "What if there are zillions of pieces of debris in the asteroid belt that we can't yet see that might damage sensitive equipment or perhaps completely destroy the spacecraft?"

Pioneer 10 didn't have the technology to detect or avoid asteroids. Hurtling through the asteroid belt, it was essentially a very expensive interplanetary bullet. Its twin spacecraft, Pioneer 11, developed to travel to Saturn, was held in storage until the scientists and engineers were certain that Pioneer 10 had survived the belt. For seven grueling months, all they could do was watch and wait. Finally, Pioneer 10 emerged on the other side of the asteroid belt completely unscathed. On April 5, 1973, Pioneer 11 was launched.

Our scientific understanding of the asteroid belt has advanced a great deal since those prehistoric days. Observations and refined theories confirm that *Star Wars*–like clumps of big asteroids do not exist, although we have discovered pairs of asteroids, one orbiting the other. Indeed, the asteroid belt is so empty that whenever we send spacecraft into it, we actually go out of our way to make sure that they do encounter asteroids, so we can study them further.

Even though astronomers have known since 1973 that the asteroid belt is nearly empty space, since 1980 almost everyone else has thought of it as chock full of matter. The reason is very simple: since the 1973 press releases announcing Pioneer 10's successful passage through the asteroid belt, there has been virtually no scientific mention of the belt in the media. However, numerous films and television shows have portrayed it as *Star Wars* did. And besides, the word "belt" in "asteroid belt" evokes the image of a solid, or nearly solid, collection of matter. In reality, asteroids are typically separated from each other by several million kilometers.

The Seasons of Your Discontent

Our practice of protecting our beliefs also creeps into perhaps the most notorious misconception in astronomy. Many, perhaps most, people have an incorrect model in their minds of what causes the seasons. This by itself doesn't make this issue notorious. That happened because the crew filming the 1988 documentary *Private Universe* asked Harvard faculty and graduating Harvard seniors to explain why the seasons change. Here these folks were, capped and gowned, feeling justifiably proud of their abilities and accomplishments, although the vast majority of them (twenty-one out of twenty-three) gave explanations of the seasons on film that were dead wrong.

WHAT DOESN'T CAUSE THE SEASONS?

The common explanation for the seasons originates in our everyday experience that the closer we are to a fire, the warmer we feel. The Sun is hot. Therefore, the closer we are to it, the warmer we should be.[1] At its closest, the Earth is about 147 million kilometers (91 million miles) from the Sun, while at its farthest, it is 150 million kilometers (93 million miles) away. So there is not much more than a 2 percent change in distance. Couldn't this account for the change in temperature, consistent with common sense?

Two things suggest otherwise. First, if the changing distance from the Earth to the Sun caused the seasons, then we would expect the entire world to have the same seasons at the same time. I wrote this on an icy cold January 26th in the dead of Maine winter. A quick call to a friend in Perth, Australia confirmed that they were enjoying a warm summer just then. Indeed, southern hemisphere seasons are exactly the opposite of those in the northern hemisphere. Second, the Earth is closest to the Sun on January 3rd each year, in the middle of winter in the northern hemisphere—one of the coldest times of the year.

If these arguments aren't convincing, we can create a model version of Earth on which the seasons are only due to the changing distance from the Earth to the Sun to see how it compares with our world. Changes in heating just from changing distance to the Sun would occur if the Earth's rotation axis were precisely perpendicular to the plane of its orbit around the Sun, called the ecliptic (figure 1.1a). Picture a skewer inserted through the Earth's rotation axis. For the real Earth, that skewer would be tilted over at an angle of 23 1/2° from the line perpendicular to the ecliptic. Now tilt the Earth so that the skewer is exactly perpendicular to that plane. If the Earth were

---

[1] Of course, if you also think that the Earth's orbit around the Sun is circular, you have to reconcile our changing distance from the Sun with a circular orbit. Perhaps the Sun is not in the center of the circular orbit. In reality, observations by Tycho Brahe (1546–1601) led Johannes Kepler (1571–1630) to discover that all planets orbit the Sun elliptically.

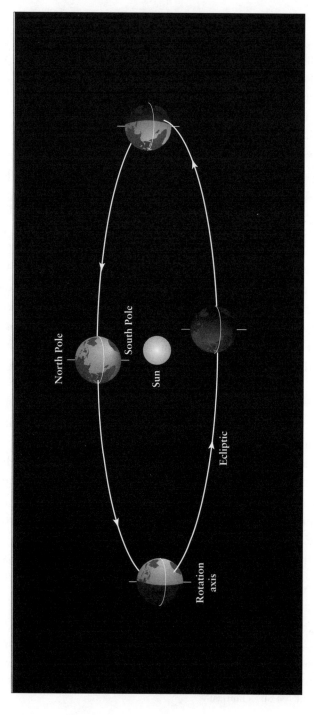

FIGURE 1.1a   Earth's orbit around the Sun as it would appear if its rotation axis were perpendicular to the plane of its orbit.

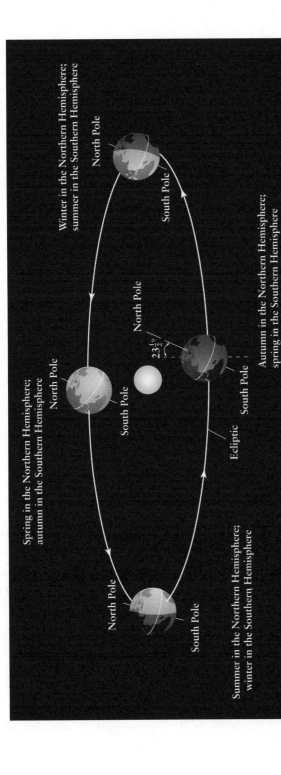

FIGURE 1.1b Earth with its axis correctly tilted 23 1/2°. Note that the axis points in essentially the same direction throughout the orbit.

oriented this way, the Sun would rise and set in the same places every day of the year as seen from any place on the planet. Furthermore, everyone would have twelve hours of daylight and twelve hours of darkness every day. Therefore, ignoring clouds, any given place would receive the same amount of heat from the Sun every day. (Places at different latitudes would receive different amounts of heat for either this or the real version of Earth, as we will explore shortly. This fact has no bearing on the present argument, however.) On an upright Earth, no change in average temperature would occur throughout the year at any fixed place *except* for the change in heat we receive from the Sun due to our changing distance from it. That is, the closer we were to the "fire," the warmer we would feel. Furthermore, both hemispheres of Earth would have the same seasons at the same time.

But how much hotter would summer be than winter? If the distance to the Sun varying by 2 percent were the only cause of changing heat on the Earth's surface, then the Earth would be 7°F (4°C) warmer in summer, when we would be closest to the Sun, than in winter, when we would be farthest from it. So assuming that the seasons occur because of our changing distance to the Sun clearly creates inconsistencies with the range of temperatures that we actually experience and with the opposite seasons in the two hemispheres.

BUILDING REALITY

Two other astronomical features could conceivably cause the seasons: the Earth's rotation rate and the tilt of its axis. The rotation rate determines how long its takes the Sun to appear to go around the Earth, that is, the length of the day. If the Earth rotated at different rates during each day or at different rates at different times of the year, then the number of daylight hours could conceivably vary. The longer the Sun is "up" during a day, the more time it has to heat that part of the Earth. Perhaps the seasons are caused by the Sun being "up" fewer hours during the winter than during the summer. We know that there are different numbers of hours of sunlight at different times of the year. Therefore, we might expect (correctly) that those months during which the number of daylight hours is longest will be the warmest. But is this due to a change in the rate at which the Earth spins?

The problem with explaining the seasons by having the Earth change its rate of rotation is that such changes are not observed; they would coincide with massive, worldwide earthquakes every few hours or few months as the planet sped up or slowed down.

This leaves the possibility that the seasons are somehow related to the observed fact that the Earth's axis is tilted compared to the ecliptic. Perhaps this tilt causes the change in the number of hours of daylight throughout the year. The angle that astronomers use to describe the Earth's tilt is the angle from a line perpendicular to the ecliptic, as shown in figure 1.1b.

Contrary to intuition, the direction in which the Earth's rotation axis points does not change throughout the year. This is because the Earth's rotation stabilizes the planet like a giant top so that the axis running between the poles always points in one direction as we orbit the Sun. As figure 1.1b shows, during half the year the Earth's northern hemisphere is tilted toward the Sun, and during the other half of the year the southern hemisphere is tilted sunward.

This explanation causes many people to leap to the obvious, but incorrect, conclusion that when the northern hemisphere is tilted toward the Sun it is closer to the Sun and therefore warmer than the southern hemisphere, hence creating the seasons. This idea breaks down when we calculate the temperature difference between the two hemispheres caused by their different distances from the Sun. That difference is only about two hundredths of a degree. This shouldn't be a big surprise, since we have just seen that changing the distance to the Sun by 3 million kilometers (1.8 million miles) changes the temperature by less than 10 degrees.

WHAT DOES CAUSE THE SEASONS?

The true cause of the seasons is a combination of two effects of the tilt of the Earth's axis: duration and intensity of sunlight. Consider the northern hemisphere when it is tilted sunward. This is the interval between March 21 (the vernal equinox) and September 21 (the autumnal equi-

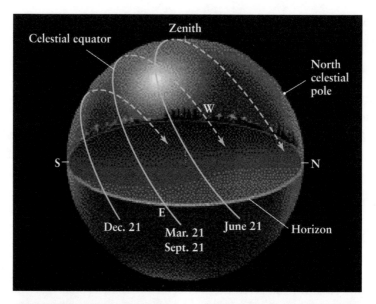

SUMMER   WINTER

FIGURE 1.2 The true cause of the seasons: the changing length of time the Sun is "up" throughout the year and the angle (and hence, intensity) of the sunlight.

nox). During that time, the Sun rises north of due east, sets north of due west, and is up more than twelve hours a day. During the other half of the year, the Sun rises south of due east, sets south of due west, and is up less than twelve hours each day. Therefore, during the period from March 21 through September 21, the duration of sunlight in the north is greater and in the south is less than during the rest of the year.

The day with the most hours of sunlight in the northern hemisphere (fewest hours of sunlight in the southern hemisphere) occurs when the North Pole is pointing most nearly toward the Sun, on or about June 22 of each year. This is called the summer solstice in the northern hemisphere. By no coincidence, this is also the day on which the Sun is highest in the sky at noon in the north. So during this half of the year, the Sun has more time to send heat onto the northern hemisphere each day, an issue discussed earlier in this chapter. Conversely, the shortest amount of daylight occurs on or about December 22 (winter solstice) of each year. The Sun is at its noontime lowest in the northern hemisphere sky (at its highest in the southern hemisphere sky) on that day, and it also heats the northern hemisphere least.

The second effect contributing to the existence of seasons is the intensity of the heat and other energies from the Sun that strike the Earth. When the Sun is high in the sky, its energy comes down more steeply, and therefore each square meter (or square yard) of the Earth receives more energy than when the Sun is low in the sky. As a result, the Earth is warmed more intensely and longer during this time than during the other six months. The amount of extra heat a hemisphere receives during the periods of longer daylight and more direct radiation is the reason the summer is warmer than the winter, and vice versa. This is summarized in figure 1.2.

Good Morning, Sunshine

While we are talking about energy from the Sun, it is interesting to explore the ideas that people have about where that energy comes from. The Sun is the source of almost all the light and heat we receive at the

Earth's surface.[2] The heat that accompanies sunlight feels reminiscent of the heat supplied by fires. It is no surprise, I think, that many people believe the energy from the Sun is created on its surface by burning gases.

Many other people conclude that the heat and light from the Sun are created by a surface covered with molten lava. Certainly, some of the images we see of the Sun in books, photographs, and movies look like the hellishly hot, glowing lava fresh from inside the Earth. Let's consider both of these possibilities.

There are many forms of burning or combustion, but generally speaking, it is a chemical reaction between two or more substances where energy is released. Often this energy is sufficient to create heat and light. Many combinations of chemicals and compounds lead to combustion. For example, many hydrocarbons, such as methane, butane, and acetylene, burn brightly when combined with oxygen and then ignited. Similarly, the slow, controlled burning of glucose in our bodies creates energy to power our actions and thoughts. Other types of burning occur when water combines with calcium carbide or when magnesium combines with oxygen, carbon dioxide, or water, among other things. The list of combustibles is quite long. For the purpose of understanding why combustion is *not* occurring on the Sun, let's consider two.

The level of the Sun from which we see its light is called the photosphere. This layer is roughly 400 kilometers or 250 miles thick. The first question, then, is: What exists in the photosphere that can burn? Chemical studies of the photosphere reveal that, by mass, it is composed of 70 percent hydrogen, 27.9 percent helium, .9 percent oxygen, .4 percent carbon, .2 percent iron, .1 percent neon, and even smaller traces of other elements.

Hydrogen is an excellent candidate for the fuel that makes the Sun

---

[2] People are always warned never to look at the Sun during an eclipse. This is sound advice. However, these stern warnings are only given during eclipses, which creates the impression for many people that it is okay to look at the Sun at other times. It is *never* safe to look directly at the Sun or at its reflection for more than a split second. The Sun is too powerful a source of energy for our eyes to cope with without extreme protection. You can buy this protection in the form of special filters sold by astronomical supply companies.

shine. After all, combining hydrogen with oxygen creates enough energy to propel the space shuttle into orbit. The amount of oxygen available to combine with hydrogen in the Sun is discouragingly low, however. Furthermore, the temperature of the photosphere is 5,800°K, or about 10,000°F. At that extreme temperature, hydrogen *doesn't* combust with oxygen, and neither do any of the other chemicals in the Sun's atmosphere. Rather, the energy associated with the heat and light from the photosphere is so great that it rips most molecules apart.

To give the combustion theory every benefit of the doubt, however, let's suppose that below its surface, the Sun were pure carbon and oxygen. Then, carbon would combine with oxygen and burn as it does in a wood or coal fire here on Earth. Even in this most optimistic of chemistries, the numbers don't add up. Assuming the Sun's entire interior were an ideal combination of carbon and oxygen, we can determine how much of these elements it would contain. Calculations in the nineteenth century revealed that the amount of carbon and oxygen possible in the Sun would only be able to burn and shine for about 8,000 years before converting entirely to carbon dioxide. Since this is much shorter than the present 4.6 billion-year life of the solar system, combustion cannot be the reason the Sun shines.

RED-HOT MAGMA[3]

Let's look at the other "obvious" possibility: that the Sun is glowing because it has a molten surface. This theory was also seriously considered in the nineteenth century. We can get a handle on whether it is correct by considering the Sun's internal composition. Recall that lava is created inside planet Earth. The rock and metal here is heated by absorbing energy from radioactive elements inside the Earth and by compression from our planet's overlying layers. Some of the interior matter has thereby become hot enough to turn molten or liquid—it is

---

[3] The phrase "red hot" evokes the image of an extremely hot object. Of all the objects that glow as a result of being hot or burning, those that are red hot are actually coolest. Those that glow with what we generally think of as the icy color blue are actually among the hottest.

magma. Upon reaching the Earth's surface, magma is called lava. So are there radioactive elements in the Sun that can convert rock inside it into magma?

Although we can't see into the Sun, we can determine its internal composition by determining its average density. Density is just the mass an object contains per unit volume. For example, the density of water is 1,000 kilograms per cubic meter (62 pounds per cubic foot). The Earth's average density is about 5,500 kilograms per cubic meter (343 pounds per cubic foot), indicating that the inside of the Earth contains many heavy elements that are denser than water. Detailed studies of lava and other surface material show that the Earth contains a great deal of iron, oxygen, silicon, magnesium, and nickel, among other dense elements.

Size is not the primary issue that determines an object's density or chemical composition. For example, here on Earth large things can have low densities while small things can have high densities. A tree floats on water while a penny sinks, because the tree is less dense than water while the penny is more dense.

Measuring the Sun's diameter and its mass, astronomers have determined that it has an average density of only 1,400 kilograms per cubic meter (87 pounds per cubic foot). This is just slightly denser than water and less than one third as dense as the Earth. What the Sun's density tells us first is that it lacks many heavy elements. This combined with how much different elements change density when they are compressed by overlying matter reveals that the interior of the Sun is a 70 percent hydrogen, 28 percent helium mix, with traces of other elements.

Knowing the Sun's overall chemical composition, we can calculate the physical state of that matter inside it. These calculations reveal that the Sun contains no solid or liquid matter at all! It does not have a "surface" in the traditional sense of the word. To paraphrase the old lady at a Bertrand Russell lecture, "It's gas all the way down."[4] The

---

[4] "A well-known scientist (some say it was Bertrand Russell) once gave a public lecture on astronomy. He described how the Earth orbits around the Sun, and how the Sun, in turn, orbits around the center of a vast collection of stars called our galaxy.

At the end of the lecture, a little old lady at the back of the room got up

Sun is a very hot gas ball right to its center, called the core, where the temperature is about 28 million°F or 15.5 million°K.

While the average density of the Sun is comparatively low, its core is staggeringly dense, 150,000 kg per cubic meter (94,000 pounds per cubic foot). That central density is created by the crushing weight of the Sun compressing the core's gas. The pressure generated by that compression is much greater than any pressure found naturally on or in the Earth. The high temperature in the Sun's core prevents the very dense combination of hydrogen and helium that exists there from liquefying, much less solidifying.

LET THE SUN SHINE IN

The correct explanation for why the Sun shines came in the 1920s when physicists, starting with Sir Arthur Eddington, realized that hydrogen atoms in the Sun's core could be forced or fused together by the pressure from the overlying gas and thereby transformed into helium. This is called thermonuclear fusion. Eddington deduced that this process releases energy by converting some of the mass of the hydrogen in the core into gamma rays.

In 1994 I attended an international conference on misconceptions in science and math at Cornell University, along with about 3,000 other educators from all around the world. One of the big sessions included a discussion of the Sun. A speaker made the obvious, and totally incorrect, point that energy cannot be created or destroyed. I watched incredulously as many people nodded and no one objected. I raised my hand and pointed out that energy *can* be created and destroyed, as long as the new energy comes from something else,

---

and said: 'What you have told is rubbish. The world is really a flat plate supported on the back of a giant tortoise.'

The scientist gave a superior smile before replying, 'What is the tortoise standing on?'

'You're very clever, young man, very clever,' said the old lady. 'But it's turtles all the way down!' " (from Stephen Hawking's *A Brief History of Time*).

namely matter. It all follows from Einstein's equation $E = mc^2$, which shows that matter can be converted to energy and vice versa.

The assembled multitude looked at me as though I were from outer space, or worse, a scientific illiterate. In their defense, I do not believe there were any other physicists or astronomers in the audience. The fact that matter can be converted into energy was proven beyond any doubt to all humanity when nuclear bombs exploded over Hiroshima and Nagasaki in 1945. The energy in those explosions was created by splitting atoms into smaller pieces, a process called nuclear fission. The Sun, on the other hand, creates energy by smashing smaller atoms together to make heavier elements (thermonuclear fusion), as just noted. Either way, energy is released and the amount of matter decreases.

To give you a feel for the enormity of this process in the Sun's core, calculations reveal that while fusing into helium, approximately 4.7 million tons of hydrogen is converted into energy every second. In other words, the Sun is losing 4.7 million tons per second. At that rate, the hydrogen in its core will allow this process to continue for a total of 10 billion years. Today the Sun is halfway through its hydrogen-fusing lifetime.

The Tale of a Comet

Besides asteroids, which are large chunks of rock and metal, and meteoroids, which are smaller pieces of the same material, comets, composed of ice and rocky debris, swarm throughout the solar system. By studying the radioactive elements in meteorites, pieces of all this space debris that have landed on Earth, we know that asteroids, meteoroids, and comets are as old as the planets. They formed 4.6 billion years ago. Mention comets and many people (incorrectly) envision balls of burning gas or molten rock with tails that streak behind them across interplanetary space. The streaks are sometimes explained by describing comets as rockets, rushing through the solar system and ejecting gases. Using this idea, many people also conclude incorrectly that tails are permanent features of comets.

As a child, I too held incorrect ideas about comets. Mine were slightly different, and came from watching the steam engines that still ran in those days. I remember watching steam shooting out the smoke stack and stopping in the air, leaving a trail behind the racing train. By analogy, I used to envision a comet as an icy object surrounded by a cloud of gas, some of which was left behind as the comet plowed through space. Let's see why these ideas are wrong.

ROCKET COMETS?

If comets are emitting gases like rockets or leaving trails behind them like steam engines, where do they get the energy to evaporate the gas? One possibility is that the comet's body generates heat as it compresses itself. However, since comets are observed to be only a few dozen miles across, the pressure inside such bodies would be insufficient to heat them enough to vaporize water, carbon dioxide (dry ice) or other volatile (easily evaporated) substances.

Radioactive elements are another possible source of heat inside a comet. Recall that radioactivity is also the primary source of heat that keeps parts of the Earth's interior molten. However, for a comet to stay molten, or at least hot enough to give off as much gas as we see in its tail, most of its mass would have to be radioactive. This is an insurmountable problem because radioactive elements become less radioactive over time as they decay into other, stable types of atoms. Since most of each comet's radioactive components would have decayed into stable atoms billions of years ago, they have too little internal heat today to create their tails, much less to be molten bodies.

Could comets receive enough heat from the Sun to be both molten and rocketlike? Because of their tails, we can observe comets much farther from the Sun than we are. Since the Sun doesn't have enough energy to make our planet's surface molten, it certainly can't melt a body farther away. Indeed, the planet Mercury, only 40 percent as far from the Sun as we are, has a solid surface. There is no known energy source that could create liquid or molten comets.

TRAILING GAS COMETS?

The other common belief is that gas given off by a comet, however it is generated, is blown behind the comet as it plows through the atmosphere between planets. This is analogous to the trail of steam and smoke left by the steam engines I watched in my youth. This model has two problems. First, where is the source of the energy that creates the gas? People often correctly approach this question by saying that perhaps the comet body, called its nucleus, is cold, rather than hot enough to be molten, but the Sun provides enough energy to evaporate volatile matter on or near its surface. This suggests, also correctly, that the comet nucleus must get close enough to the Sun for the evaporation process to begin. So far so good.

The other issue is how the tail is pushed behind the comet. The common answer, as just indicated, is that the evaporated gases are stopped by denser gas already in space, so they trail behind. Even if you haven't seen the trail of steam behind a steam engine, you have most likely seen a meteor streak across the sky. Meteors are bodies falling from space through our atmosphere and evaporating, leaving trails of glowing gas behind them. These gases are stopped by our atmosphere as the body plunges, thereby creating the appearance of a tail that trails behind the meteor and indicates the direction from which it came. But this explanation won't work for comets because the ambient gas in space between the planets is too thin to stop the evaporating gases and form a tail behind the comet nucleus. While not a perfect vacuum, interplanetary space has billions upon billions of times fewer gas molecules per volume than the air we breathe.

THE TRUE TALE OF A COMET

The notion of comets with trailing tails does have several correct elements. Observations reveal that an atmosphere of gas and dust develops around the comet's solid part as it approaches the Sun from the outer reaches of the solar system. The energy necessary to turn the comet's dust-filled ices into vapor does come from the Sun, whose ultraviolet radiation, visible light, and infrared radiation (heat) do some of the work. The comet nucleus, pounded by sunlight and high-speed particles ejected from the Sun in all directions (generally called the solar wind), frees gas and dust to form the comet's atmosphere, or coma.

By the time the comet nucleus is about 1.5 billion km (930 million miles) from the Sun, the comet begins forming a coma. It is interesting to note that while the nucleus is only a few kilometers or miles across, comas more than 100,000 km (62,000 miles) across have been observed. Amazingly, a coma can be the size of a large planet.

Often, two comet tails are visible, as many people saw from the spectacular comet Hale-Bopp (discovered in 1995 and most prominent in 1997). Studies of their chemistry reveal that one of these tails is composed of gas, while the other is composed primarily of dust-sized particles released as the surface gases of the comet evaporate. Comet tails are caused by the collisions of the gas and dust-sized particles with sunlight and the solar wind. Therefore, the cold gas model, stated earlier, is partly correct. The incorrect part is the assumption that the comet plows through gas that is just sitting there, much like the air on Earth. If that were true, the tail would trail behind the comet, which is rarely the case.

As just noted, virtually all the gas between the planets is solar wind, material that is streaming away from the Sun. The correct explanation for comet tails is that the coma gases and dust are struck by light and gas traveling *outward* from the Sun. The coma gases weigh so little that they are pushed by the Sun's light and particles straight away from the Sun (figure 1.3). This is the direction that the gas tail always points. Furthermore, we see that the gas tail[5] is frequently not uniform. It often has clumps and can even be seen to shimmer. This was an early clue to the existence of the solar wind, which does not leave the Sun smoothly. Like wind through steam or other gases on Earth, the solar wind creates ripples throughout comet gas tails.

The second comet tail is composed of dust that was freed from the nucleus by the evaporating gases. The coma dust particles are heavier,

---

[5] The gas tail is often called an ion tail, since the gas is composed primarily of nuclei surrounded by fewer electrons than usual. Such nuclei are called ions.

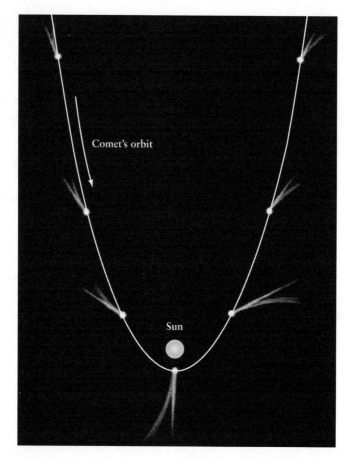

FIGURE 1.3 True direction of comet tails. The straight, gas tail always points directly away from the Sun. The curved, dust tail always arcs between the gas tail and the direction from which the comet is coming.

and while they are also pushed away from the Sun, they form an arc between the gas tail and the direction from which the comet came.

The only time a comet's tail points directly behind it is when the comet is heading straight for the Sun. Such comets have been observed. They are vaporized completely before reaching the Sun's photosphere. Most comets orbit the Sun, rather than heading straight for it. As you can see in the figure, for half the time the tails of such comets point ahead of them in their orbits.

Into the Frying Pan

While we are dealing with heat, suppose I were to tell you that the closest planets to the Sun are, from nearest to farthest, Mercury, Venus, Earth, and Mars, and ask you which one has the hottest surface. Mercury is nearly twice as close to the Sun as Venus, and nearly three times closer than Earth. Even if you had never thought about it before, it is entirely likely that you would choose Mercury as hottest. It does stand to reason. After all, Mercury is nearest the fire. However, this answer is incorrect.

If I were to stick a thermometer in space at Mercury's distance from the Sun, I would measure a temperature of 680°F or 360°C. This indicates the energy provided by the Sun at that location. Out at Venus's distance, that thermometer in space would read 375°F or 190°C. Of course, here at the Earth, the temperature is lower still, which is why we exist. As you would correctly expect, the temperature continues to decrease as we move out through the solar system. At Pluto's distance, the average temperature is -330°F or -200°C.

Let's consider now the temperatures at the planets' surfaces. Observations reveal that the surface of Mercury at noontime is a toasty 800°F or 425°C. This is hot enough to melt tin or lead. It is worth noting that Mercury has a very, very thin atmosphere compared to the Earth. Atmospheres affect temperatures. For example, on Earth, clouds prevent some heat from escaping at night, which is why mornings following cloudy nights are warmer than mornings following clear nights, all other things being equal. Since it is so thin, Mercury's atmosphere has virtually no effect on its surface temperature.

With the second planet, Venus, we have an entirely different situation. If, like Mercury and our Moon, Venus were an essentially airless world, its surface temperature would be about 450°F or 230°C, which is lower than Mercury's temperature. However, Venus is not airless. It is surrounded by so much gas, mostly carbon dioxide, that its surface air pressure is 100 times greater than the pressure of the air we breathe. At Venus's surface, your body would feel as though you were swimming 1,000 meters (3,300 feet) underwater here on Earth.

This incredibly thick blanket of carbon dioxide has a profound effect on Venus's surface temperature, raising it to 850°F or 450°C— higher than Mercury's. Let's see how. Most sunlight striking the top of Venus's atmosphere is scattered right back into space by the permanent clouds that completely cover the planet. Unlike Earth's clouds, those surrounding Venus are made up primarily of sulfuric acid. About a quarter of the sunlight does filter through to the planet's surface. As on Earth, this light is absorbed by the planet, heating it up, and as with any hot object, this heat is then radiated back outward. But the heat radiated by Venus does not get very far off the surface because the carbon dioxide atmosphere absorbs it very efficiently. As a result, the planet heats its own atmosphere, which acts like a hot blanket that keeps the surface much hotter than it would be otherwise. This is summarized in figure 1.4.

The process of trapping heat on Venus is exactly the same as that which occurs inside glass-encased things on Earth, like a greenhouse or your car on a hot summer day. Visible light passes through the glass and heats the objects inside. Since that heat cannot get back out through the glass, it heats the inside air and reheats the objects instead. This is why your car is so hot when you get into it after it's been closed up during the summer. This so-called greenhouse effect explains why Venus has the hottest surface of any planet with a solid surface in the solar system.

Ironic but true: the Earth's surface is also hotter than most of Mercury's surface. As you have experienced, the Earth cools down at night. It got to -5°F (-20°C) here in Maine last night. This drop in temperature occurs, as implied above, because the heat radiated from the surface is not being replaced with energy from the Sun. Furthermore, some of the heat emitted by the Earth is absorbed in our atmosphere (to a much lower degree than on Venus), which helps prevent the surface from cooling too much during the hours of darkness.

Now let's return to Mercury. Its rotation is so slow that a day there, from noon to noon, is 176 Earth days long. A year on Mercury is only

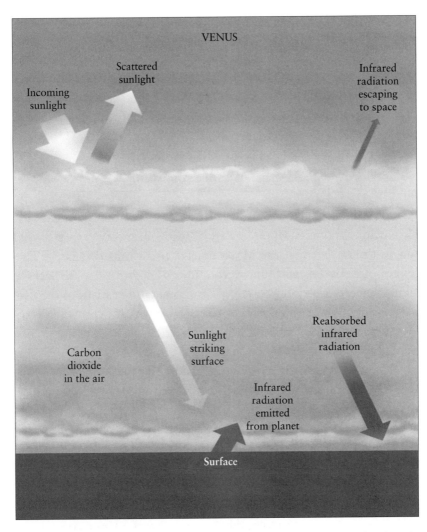

FIGURE 1.4 Greenhouse heating of Venus. Visible sunlight strikes the planet's surface and is converted into heat. Much of that heat is trapped by Venus's thick, carbon dioxide–rich atmosphere.

88 Earth days long; thus, a day on Mercury is twice as long as a year there. As a result of this incredibly slow rotation, the night side of Mercury has 88 Earth days to cool off. Since there is little atmosphere to capture any of the heat emitted by the planet, the temperature on the night side of Mercury plummets to -280°F or -175°C, making half of the closest planet to the Sun among the coldest bodies in the solar system.

Moonshine and Earthshine

As with Mercury, slow rotation, sweltering days, and frigid nights are also hallmarks of our Moon—as are a variety of incorrect beliefs. We saw in the preface that the Moon rotates as it orbits the Earth. The length of a day on the Moon is roughly 29 1/2 Earth days. It lacks a significant atmosphere. Therefore, the temperature variations are also extreme, ranging from 210°F or 100°C at noon down to -250°F or -150°C at night. Day and night have a bearing on one common area of confusion about the Moon, namely, the cause of its phases.

The most common incorrect explanation for the Moon's phases is that they are caused by Earth's shadow covering part of the Moon in the course of its orbit around Earth. A similar idea suggests that the Moon's phases are caused by shadows from other objects in space. Let's start by addressing the idea of the Earth's shadow as the cause.

MOON IN THE SHADOWS

I think everyone would accept the assertion that the Moon continually goes through a smooth cycle of phases. A quick glance at the photographs in an astronomy text or on an astronomy Web site, or a month of looking at the Moon, will convince you. The cycle goes like this: the new moon is the phase when the Moon appears either the smallest of crescents (when it passes slightly off the imaginary straight line between the Earth and the Sun) or completely dark (when it passes directly between the Earth and Sun—only on these new moons does it create a solar eclipse). As the crescent broadens, the Moon is in the waxing crescent phase; this continues for about a week until half of the side of the Moon facing us is bright. We call this the first quarter moon (because it is one quarter of the way through its cycle). The bright part of the Moon continues to widen for another week, the waxing gibbous phase (from a word that once meant "humped"), until we see a circular full moon. Then the cycle reverses: waning gibbous, third quarter, waning crescent, back to new moon.

If the phases are caused by the Earth's shadow covering different amounts of the Moon's surface on different days, then we can set up the geometry of the Sun, Moon, and Earth necessary to create the cycle. The Earth's shadow is cast when the Earth blocks sunlight. The only place the shadow exists is on the far side of the Earth from the Sun (see figure 1.5). You can see that the Sun and Moon must be on essentially opposite sides of the Earth on each day that we see less than a full moon.

This raises a problem. The Moon is only full on one day in each cycle of phases, meaning roughly one day a month. If the Earth's shadow causes the phases, then the Moon must be on the opposite side of the Earth from the Sun virtually every day of the year so that part of it can be in the Earth's shadow. But under such circumstances, the Moon would not orbit the Earth. If the Moon were positioned only on one side, the Earth's gravitational attraction would pull it straight in and onto our planet's surface. In reality, the Moon is continually falling Earthward, but because of its orbital motion it continually misses, which is a good thing. In reality, the Moon continually orbits the Earth and spends as much time on the Sun's side as it does on the side of the Earth far from the Sun. For example, take a look at a crescent moon sometime. You will always see it less than half a sky away from the Sun.

Even more challenging to the shadow–lunar phase relationship is the shape of the Earth's shadow on the Moon. For all intents and purposes, the Earth is a spherical body. When the Moon is passing through the shadow, the dark part of the Moon should always cut an arc "into" the Moon, as shown in figure 1.5. That is fine for the crescent moon, but the gibbous moon (inset) has the bright part bulging out on both sides, which is inconsistent with the idea that the shape of the Earth's shadow creates the dark part of the Moon.

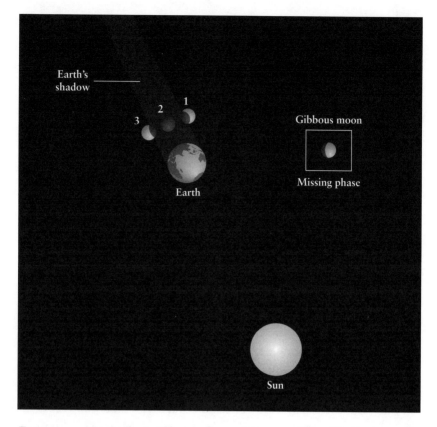

FIGURE 1.5 The scenario if lunar phases were caused by the Moon in the Earth's shadow. This model can re-create the crescent moon (1) and new moon (2), but not the gibbous moon (inset). In reality, the shape at (3) is only seen during a lunar eclipse.

Another problem with the Earth's shadow as the reason for the phases is lunar eclipses. You might well argue (correctly) that the Earth's shadow causes these eclipses. But if our shadow causes the lunar phases, why isn't there an eclipse every month? Even more confounding is the fact that we observe solar eclipses when the Moon covers the Sun. That means, of course, that the Moon must be directly between the Earth and the Sun. A few hours before or after a solar eclipse, the Moon appears as the slimmest of crescents in the sky very close to the Sun. But if the Earth's shadow caused the phases, that nearly invisible Moon would have to be almost exactly on the *opposite* side of the Earth from the Sun. These two assertions are inconsistent, so the belief that the Earth's shadow causes the phases must be wrong.

TRUE LUNAR PHASES

The actual cause of the phases of the Moon is related to the idea behind the title of a record album by Pink Floyd, *Dark Side of the Moon*. Many people believe that the Moon has a dark side, defined to be the side we never see.[6] The Moon *does* have a side we never see from Earth, called the "far side." But is it always dark?

Put yourself on the Moon for a moment. Choose a nice spot with the Earth high in the sky. Let's make it sunrise. Keep in mind that even with the Sun up, the lunar sky looks black because there is virtually no air to scatter sunlight and obscure the stars. Set up camp and watch the Earth and Sun. Because the Moon is rotating at exactly the same rate that it orbits the Earth, the Earth won't budge, no matter how many days, weeks, months, years, or decades you watch it. However, the Sun does move across the sky.

It will take the Sun just over two weeks to cross the sky. When it goes down, you will be plunged into a cold, dark night for an equal length of time. In other words, it takes the Sun one day to go from sunrise to sunrise on the Earth and nearly thirty Earth days to go from sunrise to sunrise on the Moon. But where is the Sun during the time that it is "down" in your sky on the Moon? It is illuminating the other half of the Moon under your feet, just as the Sun sheds light all around the Earth at some time each day. While half the Moon is in darkness at any time, virtually all places on it (except in some craters at its poles) receive sunlight at some time throughout the month. The dark side of the Moon is continually changing.

Even when you were in darkness on the Moon watching us, we could watch you. You stayed in the same spot, but that spot went from

[6] At the end of the album is a disclaimer stating very quietly that the Moon does not have a dark side, as just defined.

being in the bright part of Moon to being in the dark region. We on Earth see at least part of the dark side of the Moon whenever we see less than a full moon. The phases occur because we see different amounts of that dark side as the Moon orbits the Earth.

Let's see how this works, starting again at the new moon, which occurs when the Moon is between the Earth and the Sun. As the sunlit side of the Moon faces away from us, we see virtually all of the Moon's "dark side." A crescent of light appears as the sunlit side comes into view. The crescent's two points aim away from the Sun, and in this phase the Moon rises shortly after the Sun. Over roughly the next week, as the Moon orbits the Earth and moves farther away from the Sun in the sky, we see more of the sunlit side until we see half of it—the first quarter moon. From first quarter to full and then to third quarter, the Moon is more than halfway across the sky from the Sun. Indeed, when the Moon is full, it is on the opposite side of the Earth from the Sun, so that we see the entire lit side. This directly contradicts the model with the phases caused by the Earth's shadow on the Moon. There, if the full moon were opposite the Sun, it would be in our shadow. The cycle continues as shown in figure 1.6.

Eclipsed

If the Moon is on the opposite side of the Earth from the Sun every month, why isn't there an eclipse every month? There would be if the Moon orbited the Earth in exactly the same plane as the Earth orbits the Sun (the ecliptic). In that case, every time the Moon was about to be full, it would slide into the Earth's shadow, creating a lunar eclipse. Similarly, every time the Moon was about to be new, it would pass exactly between the Earth and the Sun, blocking some sunlight and thereby creating a solar eclipse.

In reality, the Moon's orbit is tilted by about five degrees to the ecliptic, so that usually it is slightly above or below the Earth's shadow when it is full and the Earth is slightly above or below its shadow when it is new. Eclipses only occur when the Moon is crossing the ecliptic at those two phases.

Phases of the Moon

C
First quarter
Moon

D
Waxing gibbous
Moon

6 PM

B
Waxing crescent
Moon

Light from Sun

E
Full Moon

Midnight —

— Noon

A
New Moon

Waning gibbous
Moon

6 AM

Waning crescent
Moon

F

Third quarter
Moon

H
Light from Sun

G

FIGURE 1.6 Correct explanation for the phases of the Moon. Except at full
moon, we see part of the Moon's dark side. The amount we see determines
the phase. The drawings of the Earth and the Moon are as they would appear
from space. The photographs are views of the Moon seen from the Earth.
Note that the same craters and other lunar features are visible throughout the
cycle of phases.

If you have looked at the Moon often enough, you have probably noticed that the dark side is often slightly visible along with the bright side. Some people attribute this to light emitted by the Moon itself from a glowing molten surface, or to starlight or sunlight scattered by the Moon's atmosphere. In fact, the stars are too dim to make the dark side of the Moon visible from Earth. Its surface is solid and it does not glow on its own. Its atmosphere is almost nonexistent, so it can't scatter light. The correct reason the dark side is slightly visible is that some of the sunlight striking the Earth scatters back into space and strikes the Moon's surface. Some of this light, called Earthshine, reflects off the dark side facing us and comes back.

The Tides of March

The other common area of confusion about the Moon is its relationship to the tides on Earth. Among the incorrect beliefs about tides are the following: they are due to winds on the ocean; the Moon does not cause the tides; tides are just due to the Moon; the height of high tide doesn't change throughout the year; there is only one high tide and one low tide on the Earth at any time; there is a high tide at the point directly between the Earth and the Moon; and the spring tide only occurs in the spring of each year.

Anyone who has spent time at the ocean quickly becomes aware of the tides. I first learned about them as a child living in New London, Connecticut. We often went to Ocean Beach during the summer, and I built countless sand castles, many of which were washed away by the rising tide. The ocean rises. It settles. We observe a predictable cycle of tides. Observations such as these will lead us to their correct explanation.

Do Winds Cause the Tides?

If the tides were due to winds on the ocean, then when the winds die, the water should stop changing height. The rising and lowering of the ocean is almost always imperceptibly slow. If you go down to the beach on a calm day and fail to see any change in the ocean height while you watch it, you might well conclude that the lack of wind has stopped the tides. You can change that belief by setting up a beach chair and umbrella and watching the calm waters for a few hours. They will change height regardless of the wind strength. Also, the fact that the tides cycle reliably should put the wind-tide relationship in doubt. If you believe that winds cause tides, then if the wind is calm for a day, you shouldn't see normal cycles of tides. But you will.

Observations often help correct beliefs about the natural world. The problem is that we often don't know our beliefs are erroneous, so we don't bother to test them with observations. Worse, incorrectly interpreting observations can lead to additional incorrect beliefs. I believe the major source of confusion about winds and tides comes from the reality that winds do cause ocean waves. Go to the beach on a windy day and you will see waves run farther up the beach than on a calm day. This occurs because higher winds generate higher waves.

To be fair, extreme winds can actually change the heights of the tides. During hurricanes the height of the high tide increases due to lower air pressure above the water. This change is called a storm surge, but it only occurs noticeably during severe storms.

DOES THE MOON ALONE CAUSE THE TIDES?

Sailors have known for millennia that the height of the tide is related to the location of the Moon. For example, high tide occurs just after the Moon passes its highest point in the sky each day. This strongly suggests that there is a physical relationship between the Moon and the tides. If the Moon were the only cause of the tides, then the tides would be the same height every time the Moon was at the same place over the Earth. In other words, each day when the Moon was just past being highest in our sky, we would have a high tide of the same height as on all previous days. However, observations reveal that the height of the high tides varies throughout the cycle of lunar phases.

Well, you might argue, then perhaps the Moon's orbit around the Earth isn't circular. If that were the case, when the Moon is closer to the Earth, the Moon's gravity would be stronger, making high tides higher than they would be when the Moon is farther away. Indeed, observations reveal that the Moon's orbit is elliptical. Newton's law of gravitation explains that the closer two bodies are, the greater the tidal forces between them. When the Moon is closest to the Earth and high in our sky, the high tides *are* often higher than average.

However, further observations reveal that sometimes when the Moon is closest to us, the high tides are lower than usual, and when the Moon is farthest from the Earth, the high tides can be even higher. These last results are at odds with the belief that the Moon is the only cause of the tides.

ARE THERE ONLY ONE HIGH AND ONE LOW TIDE ON EARTH AT ANY TIME?

I went to Mount Desert Island on the coast of Maine to see the tides for myself. I arrived shortly before the Moon came up on the eastern horizon one morning[7] and noted the height of the tide on one of the pilings in Southwest Harbor. I stayed there for about an hour, drinking coffee and watching the harbor activity. During that time the tide first went down slightly and then began climbing as the Moon rose. It was brisk that morning, but the sky was clear and the day looked promising.

I spent the rest of the morning walking on various carriage trails in Acadia National Park, enjoying the breathtaking landscapes. As lunchtime neared, I returned to Southwest Harbor and had a clam basket and side order of onion rings at Beal's Lobster Pier. As the Moon passed its highest point,[8] I interrupted my lunch several times to again measure the height of the tide on the same piling. The tide rose to its highest level and then started back down during that period. The time interval between the low point that morning and the high point was roughly 6 hours and 15 minutes.

---

[7] Keep in mind that the Moon rises at all different times of day and night, depending on the phase it is in.

[8] Contrary to common belief, the Moon is visible during the day. Indeed, it is visible as often during the day as during the night.

I spent the afternoon in Acadia National Park, going up Cadillac Mountain, sitting by Thunder Hole, lying on Sand Beach. The day passed all too quickly, and at around five o'clock, I went back to Beal's for a lobster dinner and to again watch the tide, which was lowest at around 6 P.M., about 6 hours and 15 minutes after high tide. By the time I left the island at about 8 P.M., the tide was rising.

But wait a minute. I just experienced two low tides in one day and, all things being equal, there should be a high tide before the next low tide the following morning. That means two low tides and two high tides each day, not one.

There is another way of getting to this point: at the same time I saw the Moon rising in the east that morning, someone on the other side of the world was seeing it setting in the west. Since I experienced a low tide when the Moon was rising, it seems plausible that there would be a low tide on the opposite side of the Earth from me at the same time. This is consistent with the fact that I witnessed another low tide as the Moon set. Likewise, when I experience a high tide just after the time the Moon is highest, someone on the opposite side of the world must also experience one. That means there is a high tide when the Moon is high in the sky and one when it is high over the opposite side of the Earth.

"Springen"

Many people believe that the spring tide only occurs in the spring. The fact is, the spring tide occurs roughly every two weeks throughout the year. The name doesn't derive from the season. Rather, it comes from the German word *springen*, which means "to spring up." Spring tides are when the water "springs up" the most—the days of the highest high tides and lowest low tides.

Causes of Tides

Tides result from centrifugal and gravitational forces acting on the Earth. By far the most important sources of these tide-generating forces are our Moon and the Sun. Consider the case of the Moon. It and the Earth actually orbit a common point, called their center of

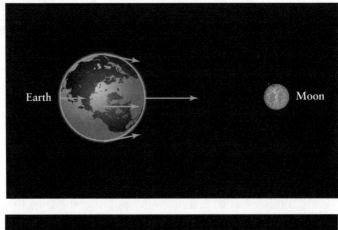

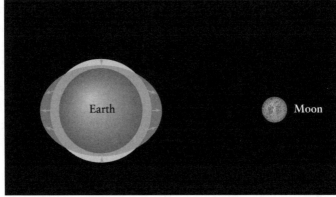

FIGURE I.7 Tides created by the moon. The arrows (top) show the strength of the gravitational force from the Moon at different places on the Earth. The resulting differences in force creates the tides, as shown below.

mass, located about 1,000 miles below the Earth's surface in a line between the Earth and the Moon. Like two dancers holding each other and whirling around, this orbital motion creates a centrifugal force on Earth's oceans, acting away from the Moon. At the same time, the Moon's gravitational pull attracts different parts of the Earth toward it by different amounts (figure 1.7). Combining the outward centrifugal force with the inward gravitational force, the tides occur as ocean water is dragged toward or away from the Moon.

There is symmetry in these interactions. Different ocean water drains from halfway around the globe either toward the Moon or away from it, simultaneously creating equivalent high tides on two sides of the Earth (and two low tides in between them). This is consistent with my observations on Mount Desert Island that there are two high tides and two low tides each day. The high tides occur when the Moon is high in the sky or when it is beneath our feet. The low tides occur when the water near us is being pulled along the Earth's surface toward or away from the Moon—when the Moon is near the horizon.

The Sun's gravity also has a significant effect on Earth's tides. After all, the Sun is 27 million times more massive than our Moon. However, the greater distance between the Sun and the Earth—nearly 400,000 times the distance between the Moon and the Earth—significantly diminishes the Sun's tidal effects here. Combining the effects of the Sun's greater mass and greater distance leads to the result that the Moon creates tides about twice as high as those created by the Sun.

We are now in a position to understand why the height of the high tide changes throughout the cycle of lunar phases. When the Earth, Sun, and Moon are in a straight line, at either new moon or full moon, the Sun and Moon pull on the oceans in the same direction. Note that they do so even when on the opposite sides of the Earth because of the symmetry of tides created by each body. At new and full moon, the range from high tide to low tide is especially high. These are the spring tides.

When the Sun and Moon are pulling on the Earth in perpendicular directions, say when the Moon is just rising at noontime (when the Sun is highest in our sky), the tides they create partially cancel each other out. This occurs because the Sun and Moon are pulling the oceans in competing directions at the same time. It is most noticeable at the first and third quarter lunar phases. Tides on these days, when the range between high and low tide is especially low, are called neap tides. We can now see how the changing distance from the Moon to the Earth is only part of the tidal story: when the Moon is closest to us at neap tide, the tidal range is still lower than when the Moon is farthest from us at spring tide.

Finally, let's put the Earth's rotation back in the picture. The Earth is spinning eastward roughly thirty times faster than the Moon is orbiting around us. This motion pulls the high tide from beneath the Moon. The Moon's gravity pulls the water in the high tide closest to the Moon back toward it. This means that the high tide flows westward over the Earth's surface due to the Moon's attraction. The high tide is therefore not directly under the Moon. Furthermore, the water is stopped by the islands and continents. This process does two things: first, the high tide nearest the Moon pulls the Moon ahead in its orbit (figure 1.8). This gives the Moon extra energy, causing it to spiral away from the Earth. It is presently receding at a rate of nearly two inches (four centimeters) a year. Second, the westward-moving high tides push on the eastward-spinning continents, thereby slowing the Earth down. Based on this physical and geological evidence, geologists have determined that when the Earth formed, it was spinning five to six times faster than it is now. The day was originally somewhere between four and six hours long. The Earth's rotation rate is presently slowing down by about a thousandth of a second per century.

Learning this often generates the question, Will the Moon ever leave the Earth completely? It will not, because the Moon moves away in proportion to the Earth's slowing rotation rate. Eventually, perhaps some 20 billion years from now, the Earth will be spinning at the same rate as the Moon is orbiting it. Thereafter, the Moon will remain at a fixed distance from Earth with a high tide directly between the centers of the two bodies. The Moon will then appear fixed over one side of the Earth, never to be seen on the other side. This extrapolation into the future is moot, however, because the Sun will have stopped shining long before this and, in all likelihood, will have swallowed the Earth and Moon in the process.

Fifty Commonly Cited Incorrect Beliefs About Astronomy

On every topic in astronomy, indeed on nearly every topic of science, people hold myriad incorrect beliefs. The common, incorrect ideas about astronomy I have explored in this chapter literally describe the tip of the iceberg. Over the past decade, my introductory college astronomy students provided me with lists of incorrect beliefs they had

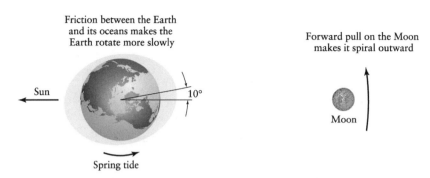

FIGURE 1.8 The Earth's rapid rotation creates off-center tides. The high tide nearest the Moon leads it by 10°. The bulge pulls the Moon ahead, causing it to spiral away from the Earth.

prior to taking the course. They also told me where they learned those ideas. The incentive for them to provide this information was, of course, extra credit and the guarantee that what they wrote would never be associated with their names.

As a result, I compiled a list of more than 1,560 different incorrect beliefs. Please note that this doesn't mean that my students are not intelligent. As we will explore in detail in the next chapter, we all develop incorrect ideas about the natural world. It is unavoidable. Just listing all of them would take fifty pages. Presenting explanations for each wrong belief, together with the correct science, as I've done for several in this chapter, would take hundreds of pages. Instead of including them here, I have put the entire list on the Web at http://www.umephy.maine.edu/ncomins/. You can also tally there the incorrect beliefs that you had, add new ones, provide correct explanations about phenomena on the list, and suggest links to correct information.

Below are the fifty most commonly cited incorrect astronomical beliefs provided to me by the students. Keep in mind that these are not necessarily the most common ones that exist—they are just the ones that come to mind most readily when people are asked about the astronomical ideas they had but have since learned were erroneous.

1. Pluto is always the farthest planet from the Sun.
2. Stars really twinkle.
3. The Sun primarily emits yellow light.
4. There are twelve zodiac constellations.
5. The constellations are only the stars making the patterns.
6. Saturn is the only planet with rings.
7. Seasons depend on the distance between the Earth and the Sun.
8. Polaris, the North Star, is the brightest star in the night sky.
9. The asteroid belt is densely packed, as in *Star Wars*.
10. Giant planets (Jupiter, Saturn, Uranus, and Neptune) have solid surfaces.
11. We see all sides of the Moon from the Earth.
12. Black holes are huge vacuum cleaners in space, sucking everything in.
13. All planets have prograde rotation (they spin in the same direction as they orbit the Sun).
14. The only function of a telescope is to magnify.
15. A shooting star is actually a star falling through the sky.
16. Comet tails are always behind the comet.
17. Black holes last forever.
18. All moons are spherical.
19. Only the Moon causes tides.
20. Ours is the only moon.
21. The Sun is a unique object, not a star.
22. Pulsars are pulsating stars.
23. Spring tide only occurs in the spring.
24. Saturn's rings are solid.
25. The Sun will last forever.
26. The Sun shines by burning gas or from molten lava.
27. There is a permanently dark side of the Moon.
28. Black holes are empty space or holes in space.
29. The Sun doesn't rotate.
30. The Sun is solid.
31. All stars are yellow.
32. The Moon is not changing distance from the Earth.

33. Stars last forever.
34. There are many stars in the solar system.
35. The Sun always rises directly in the east.
36. Meteors, meteorites, meteoroids, asteroids, and comets are all the same things.
37. The galaxy, the solar system, and the universe are all the same thing.
38. The Big Dipper and the Little Dipper are constellations.
39. Mercury is hot everywhere on its surface.
40. Gravity is the strongest force in the universe.
41. Once the ozone is gone, it's gone forever.
42. There are only a few galaxies in the universe.
43. The universe is static or unchanging.
44. The center of the Moon's core is at its geometric center.
45. Jupiter's Great Red Spot is some kind of surface feature.
46. All planetary orbits are circular.
47. High tide only occurs between the Earth and the Moon.
48. Comets are burning and giving off gas as their tails.
49. The Earth is at the center of the universe.
50. All galaxies are spiral-shaped.

I want to explore what I believe are the major issues about our numerous incorrect beliefs, using many of them as examples. To start, I will ask: When and where do we get these ideas? How is it that we function so well while harboring so many wrong ideas about nature? Why are these beliefs so hard to correct? Indeed, can they be corrected, and if so, how? Can we avoid formulating new wrong beliefs?

Our incorrect beliefs begin accumulating when we are infants. Cartoons are an early source of them. Another early problem we all face is as old as humankind itself. Words are essential for so much of our communication, yet many words mean different things to different people. If two people discuss something, say the solar system, and each interprets the words differently, then they are unlikely to understand each other. Words and their multiple uses are a major source of misconceptions. What is more ironic, one of the misleading words is the word "misconception" itself.

2

Blame It on Someone Else

EXTERNAL ORIGINS OF INCORRECT BELIEFS

Cartoons and Science Fiction

'TOON TOWN

We begin learning about the natural world as children, initially developing expectations about what will happen based on our experiences. An early understanding of the laws of physics develops from our constant struggle against the force of gravity. By the time children are one year old, they have firsthand experience with gravity's effects as they learn to walk. A more abstract example is that infants apparently believe that when someone disappears from view, the person no longer exists.[1] Experience helps them replace that belief with one of "object permanence."

Of course, children have a relatively limited range of direct experiences with the natural world. Much of their understanding of how it works comes from television. The problem is that many of the cartoons that young children see don't portray nature accurately. As you probably remember from your own childhood, youngsters develop several incorrect beliefs about gravity from such cartoons as the Roadrunner. The reason Wile E. Coyote momentarily stops moving in midair after going over the cliff is, of course, to give the character and the young viewers time to realize the danger of the situation—he inevitably looks down and then looks at the audience. To their credit, some channels now explain to children that this is an unreal, cartoon effect they call "delayed gravity."

Another incorrect belief that these cartoons generate in children is that after an object goes off the edge of something, it stops moving forward and therefore falls straight down. When my younger son Joshua was seven, I took a penny and asked him what path it would follow if I slid it off the kitchen table. He showed it making a right angle in the air and then falling straight down.

In reality, anything moving straight off a horizontal surface begins to fall immediately, but it also keeps moving forward. In fact, as things fall they continue to move forward at the same speed they had just before they began falling. As a result, the objects follow a path shaped like an arc, technically called a parabola. You can demonstrate these things by sliding a penny off a smooth surface.[2]

Another example of the effects of cartoons on our perception of the natural world is Hanna-Barbera's *The Flintstones*. It creates the impression, and most children therefore believe, that dinosaurs and humans coexisted on Earth. Indeed, it had a great impact on me as a child, since I was in one of the test audiences that rated the first *Flint-*

---

[1] It's an extreme example of the more grown-up attitude of "out of sight, out of mind."

[2] Don't use a pet hamster for this—it will not stop in the air so it can be grabbed before it falls, as at least one young person I know found out from experience.

stones television pilot way back in the summer of 1960. I was in college before I learned that roughly 61 million years passed between the time the last dinosaur walked the Earth and when our earliest hominid ancestors looked up at the night sky.

SCIENCE FICTION

As children age, cartoons give way to other forms of fantasy entertainment. For some, science fiction becomes a lifelong passion—and a lifelong source of incorrect beliefs about the natural world.

"But a lot of stuff you read or see in science fiction eventually becomes real."

I agree. Space travel and submarines, along with laser beams, wrist radios, and a variety of other current high technologies were all proposed in advance in science fiction books and movies. My focus is on those concepts in science fiction, like antigravity devices and faster-than-light travel, that violate tried and true laws of physics.

"But what if those laws of physics are proven wrong?"

The word "wrong" is the key here. Scientists have developed physical laws that describe how various aspects of nature work. Many of these laws have led to the development of innumerable high-tech devices, such as electronics, lasers, aircraft, rockets, and magnetic levitation trains, among many others. The accuracy of such laws in predicting how nature behaves and in leading to successful products are our measures of the laws' validity. When they are pushed into new realms, as when the laws of physics that describe the motion of a truck are used to describe the motion of small clusters of atoms, they eventually fail. The point is, however, that our successful physical laws fail at some limits of their validity, not at their cores. Instead of calling them wrong, I would say that most laws of physics are incomplete in describing how things work. Put another way, physical laws are described by mathematical equations, some of which are more accurate representations of reality than others. The less complete laws predict phenomena that experiments show are not possible—but these results are almost always at the periphery of our scientific understanding of nature.

I believe that such things as antigravity devices and faster-than-light travel should be classified as incorrect, since there is no experimental or theoretical evidence whatsoever today that they are possible and much evidence that they are not. If experiments prove otherwise, I will be the first to admit I'm wrong. Until then, belief in anything that violates experimentally established laws of nature is to me incorrect.

Don't get me wrong. I enjoy a good science fiction story as much as anyone. Part of the process of "getting into" science fiction is suspending our disbelief in things that cannot happen or are false or impossible. I'm all for suspending my disbelief for the prospect of good entertainment. The issue is to avoid incorporating any of the invalid, fictional ideas you encounter during such times into your understanding of the real world. This is the hard part—separating fact from fiction, especially if you don't know that the fiction you are watching or reading *is* fiction.

We are especially susceptible to believing incorrect ideas when we are being entertained. This occurs because when we suspend our disbelief, we generally don't do so selectively. To make the most of the experience, we enter the world created by the writer, actors, or virtual reality simulations without reservation. Once inside these alternative worlds, we don't have a mental firewall that keeps evaluating and reminding us what is fact and what is fantasy. At best, that comes later.

Perhaps the most insidious problem in letting our guard down is when we encounter situations or events that seem scientifically plausible but are inaccurate. The asteroid belt image created in *The Empire Strikes Back* is an excellent example of that. If you don't think about the effects of gravity, it certainly seems plausible. Since the image fits well with our concept of a "belt" as a more or less solid band of matter, we let it drive our vision of the real-life asteroid belt.

When we suspend disbelief uncritically, we are taken even further from the reality dictated by the laws of nature than just the fiction in science fiction. This suspension opens whole new realms to be incorporated into our understanding of the world. By allowing concepts inconsistent with known science to influence us, we are more likely to draw incorrect conclusions about nature.

Pandora's Box

In the summer of 1997 I attended another conference on misconceptions at Cornell. During the last session, Professor Joel Mintzes, a distinguished biology educator, gave a talk on the evolution of his beliefs about astronomy from his childhood through the present. He did this with a set of diagrams showing various astronomical beliefs he had at different ages. Not surprisingly, many of the beliefs he held as a child and adolescent were incorrect, although the number of them decreased with age. However, even the diagram showing his current beliefs contained substantial errors (figure 2.1).

I tell this story not to demean Joel but to emphasize that even highly educated professional scientists and educators have numerous incorrect beliefs about the natural world. Indeed, he is remarkable because when I e-mailed him a request for copies of those diagrams and permission to write about them, he wrote back, saying, " 'You are most welcome to my concept maps (any or all of them) and you may use them with or without attribution, as you wish. One of the points I have tried to make over the past 25 years is that there should be no shame associated with conceptual errors, either within or outside of one's field of 'expertise.' "

While I may have fewer incorrect beliefs about astronomy than you do, I probably have as many about economics, sociology, the law, paleontology, botany, and other fields in which I am not an expert. So, beyond childhood, where do our incorrect beliefs come from? Their origins fall into three rough categories. In this chapter I explore external origins; in the next chapter I consider origins due primarily to our experience and reasoning and to a combination of both external and experiential sources, including that most insidious cause of all: common sense.

Words Get in the Way

TOWER OF BABEL

For many years I used the word "misconception" to describe incorrect beliefs. Let me invite you to think about your definition of that

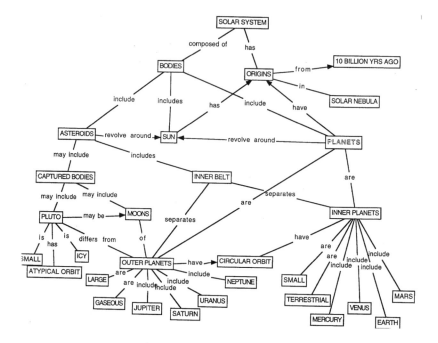

FIGURE 2.1 Diagram of astronomy beliefs of a Ph.D. biology educator done at age thirty-nine. Besides errors in the age of the solar system and the shapes of the orbits, omissions include comets and meteoroids. Meaning of "captured bodies" is unclear, but suggests an incorrect belief since astronomers believe that asteroids were formed with the rest of the solar system and that Pluto is fundamentally different from the asteroids.

word before I give you mine. I wrote papers about misconceptions and gave talks about them. Then came that conference on misconceptions in science and mathematics at Cornell in 1994. Listening to presentations and talking to hundreds of other people who are interested in how we learn, I quickly became very confused. Therefore, I asked more than fifty researchers to define the word "misconception." The range of definitions was disconcerting in the extreme. It included the following:

- any belief that is untrue
- only incorrect beliefs that are deep-seated in our minds
- all beliefs incompatible with currently accepted scientific beliefs
- all deep-seated beliefs incompatible with currently accepted scientific beliefs
- any belief that we derive by misinterpreting information
- any belief that we derive by misinterpreting our senses
- incorrect explanations derived by children to explain the natural world
- an incorrect mental construct
- a conception or belief that produces a systematic pattern of errors
- a mistaken idea
- a misunderstanding
- a false opinion
- a misguided view

This list is far from complete, and you can add your own definition, if you wish, on our Web site. While there was some overlap between definitions, no two people gave identical ones. So there we were, describing research about misconceptions and how to replace them with scientifically correct knowledge, when none of us was quite sure of the way the rest of us were using certain important words. To make matters worse, I never heard speakers give their own definitions of the word "misconception," so I was left trying to figure out which version was being discussed.

The implications are staggering. People can communicate in part because we share a basic understanding of what the words we use mean. You and I could talk about cars because we both have consistent ideas of what cars are. More abstract concepts, like "community," require more careful treatment, since our ideas of what a community is could be radically different. I was reminded of the tower of Babel at the Cornell meeting, where people were talking to each other about "misconceptions" and often misunderstanding what others were saying because they each interpreted that word differently.

First Come, First Used

Some people who study how we learn and how to teach believe that the word "misconception" has a negative meaning. They are concerned that if people are told that they have misconceptions, they will get defensive because it implies that they lack intelligence or education. The psychology of keeping people from getting defensive about their misconceived ideas has motivated the introduction of other words and expressions to replace "misconception," such as "prior beliefs," "prior perceptions," "naive beliefs," "naive theories," "naive concepts," "preconceptions," "alternative beliefs," "alternative conceptions," "alternative frameworks," and others. These words and expressions do not come with as many pejorative connotations[3] as does "misconceptions."

It is my experience that the first definition or explanation I learn about anything is the one I use unless unbearably compelling evidence forces me to change. That is the problem with using common words in specialized situations and why new words are more desirable. A new word comes with a new definition, and if everyone starts out with the same definition, that word (or phrase) is much more useful in communication than words that already carry multiple meanings. Unfortunately, introducing new words is extremely hard to do successfully, and there really is no untarnished preexisting word or expression to replace "misconceptions."

I define a misconception as any deeply held belief that is inconsistent with currently accepted scientific concepts. Deeply held beliefs are distinctly different from superficial ones, such as details we might memorize for an exam and then promptly forget. Deeply held beliefs are those we have incorporated into our understanding of the world and use in a variety of different contexts.

---

[3] To show you how hard the issue of language is, consider this: because these phrases are written using standard words, even just reading the words has started you thinking about their common meanings. Therefore, you would actually have a prior belief about what each expression means before you discover its technical definition in learning theory.

Because they are well embedded in our understanding of nature, we often use science-related misconceptions to try to understand other things we encounter for the first time. When our original ideas are incorrect, the conclusions we draw using them are often incorrect too. This is one of the great problems with misconceptions—they propagate. Many people hold and use deep-seated beliefs that are wrong, such as that the positions of the heavenly bodies at the time of your birth can affect your personality; that it always takes some force acting on a moving object to keep it in motion; that heavier objects fall faster than lighter ones; that plants grow primarily by converting soil into organic material; that objects as massive as trains can stop as quickly as objects as light as cars; or that the harder you push on something the more work you are necessarily doing. These are all misconceptions.

Conversely, simple incorrect facts you get and accept, usually from external sources like books or friends, are not misconceptions. They are incorrect beliefs, but not deep seated. Examples of superficial incorrect beliefs include learning the wrong number of moons orbiting Jupiter or misidentifying the make of a car you see or getting the order of the planets outward from the Sun wrong.

It would be impossible to write a book about misconceptions that satisfies everybody's definition of that word. Therefore, I will often refer to incorrect beliefs when I want to discuss both misconceptions and less deep-seated ideas inconsistent with scientific knowledge. Since such knowledge is the foundation for our understanding of the cosmos, it is also important that we understand what the word "science" means.

FISH AND FOWL

Science is actually two things. First, it is a body of knowledge gathered from experiments and observations, together with mathematical explanations of that information. For example, geologists discovered evidence of liquid water on the early Earth that indicates that the Sun has been shining for more than 4.5 billion years. Physicists then derived an explanation for the Sun's ability to glow for this length of time—

nuclear fusion in its core. Generalizing from this explanation, we believe that the other stars in the universe also shine because of nuclear fusion deep in their interiors.

Second, science is a process. In the typical scheme of things, people begin by making observations or doing experiments.[4] They then often create a variety of hypotheses to explain the observations or experimental results. Hypotheses are provisional or working ideas that are then tested. Those that prove valid become scientific theories, or scientific laws, as they used to be called. Both hypotheses and scientific theories must be mathematical so that they can be used to make quantitative predictions. These predictions are tested by doing further experiments or making further observations. Theories that prove inconsistent with this work are modified if possible, or else they are discarded. The theory that most accurately explains observations and experimental results and predicts the results of further observations or experiments is adopted as the correct explanation.

In the case of a tie between two or more scientific theories making equally accurate predictions or providing equally reasonable explanations, the simplest theory is usually chosen. This is called the principle of Occam's razor, after the Franciscan monk William of Occam, who first espoused it in the early fourteenth century. He wrote, *"Pluralitas non est ponenda sine necessitate,"* which translates into "plurality should not be posited without necessity." In other words, choose the simplest scientific theory to explain whatever it is you are studying. Scientists follow this concept scrupulously because it works—among competing theories that explain the same experiments or observations, we always choose the simplest one.

Consider this example of Occam's razor from the sixteenth century. In 1543 Copernicus published his *De Revolutionibus Orbium Coelestium* (The revolution of the celestial spheres) in which he proposed that the Earth and planets have circular orbits around the Sun. Throughout human history, most people[5] believed that everything in

---

[4] A counterexample is the Big Bang theory of the expanding universe. In that case, the equations preceded the observations.

the sky had circular orbits around the Earth. This was codified by Ptolemy in the second century A.D. His reasoning was simple: look at the night sky and you see the apparent motion of everything circling the Earth. However, when the Ptolemaic system was compared to detailed observations, corrections had to be made in the form of little circular motions of the planets and Sun, called epicycles. Epicycles are superimposed on their grand circular orbits around the Earth. As observations got better, more and more corrections had to be added to the Earth-centered model of the universe.

Copernicus used his Sun-centered model to make predictions about the positions of the planets. His results were *no better* than the predictions of the Ptolemaic Earth-centered model! However, the Sun-centered model was much simpler, not requiring a vast array of corrections, while giving equally accurate results. The predictions of this early Sun-centered model were not perfect, but the model was simpler than the Earth-centered model. This, and this alone, was the appeal of the Copernican model until 1609, when Johannes Kepler began publishing his observation-based laws that concluded that planets have elliptical orbits around the Sun. This correction to Copernicus's theory was all that was needed to predict the positions of all the planets as precisely as we can observe them, even today (with the exception of the orbit of Mercury). So the Copernican revolution was based on Occam's razor.

Scientific Seers

The best theories also make predictions about things beyond the observations they were initially designed to explain. If these predictions are borne out, we keep expanding the range of validity of the theory. When the predictions are shown to be incorrect, we modify the theory until it works or, failing that, acknowledge the theory's limited range of validity.

For example, Kepler's three laws predict the orbital positions of

---

5 The most famous exception is Aristarchus of Samos, a third-century B.C. philosopher, who believed that everything orbited the Sun.

the planets but don't explain *why* the planets follow these orbits, and they are not useful for predicting the motion of things like soccer balls flying through the air here on Earth. The first viable explanation of why planets have elliptical orbits was derived by Sir Isaac Newton and published in his *Principia* in 1686. Still used today, his law of gravitation also correctly predicts the path of a soccer ball or any other object thrown through the air or dropped. It even accurately predicts the path of the space shuttle in orbit around the Earth, among thousands of other things. Kepler's laws are consequences of the laws of gravitation and conservation of momentum and energy derived by Newton. Therefore, Newton's laws have a much wider range of application than Kepler's.

However, even Newton's law of gravitation has limits. For example, it makes incorrect predictions about the orbit of Mercury around the Sun. The explanation of that planet's orbit is embodied in Einstein's theory of general relativity, which makes accurate predictions for the positions of all the planets, including Mercury. General relativity makes the same predictions about motion on Earth as does Newton's law of gravitation. The cost, however, of the more encompassing general relativity concept is the need for much, much more complex mathematics. Newton's law of gravitation, while not as comprehensive, is accurate and simple enough for application in everyday life, and that is why we still use it.

WORDS AND ACTIONS

I hope I have demonstrated that since language, even the word "science," is often imprecise, we don't always understand what others are saying. Indeed, we often misunderstand what other people tell us about matters in absolutely every realm of our lives. We overcome imprecision in our everyday language by taking into account the context in which words are used, by adding more words to clarify matters, or by asking questions. Then, when all is said, we watch actions to determine if the words are consistent with them.

In my introductory astronomy class, I encourage questions and receive perhaps a dozen each class. But consider what is happening. A student is asking about something from his or her own perspective, often using words completely differently than I use them. I was once asked by a student, "When did the solar system form?"

My answer was, "Four point six billion years ago."

He immediately objected, saying that he had recently heard on the news about stars in the solar system that were 12 billion years old, older than the purported age of the universe.

At issue was the definition of "solar system." To the student it was the Sun, planets, moons, and all the stars in the sky, a common belief. He was confusing what astronomers call our galaxy (the Milky Way), which contains hundreds of billions of stars, with what we call the solar system—just the Sun and everything that orbits it. For many years now, I have tried to give astronomical definitions of terms I use in answering such questions.

A related issue is that we often misunderstand questions and there-fore answer different ones than were intended. I was once asked, "When were the craters on the Moon made?"

I answered that most of them were made during the first billion and a half years that the solar system existed.

"No," the student said, "I mean, what made them? Were they made when volcanos erupted on the Moon, like craters on Earth?"

I then explained that as far as has been determined, all lunar craters were made by impacts of space debris on the Moon. I have taken to asking, "Did I answer your question?" at the end of every explanation I give.

SPACE DEBRIS KEEPS FALLING ON MY HEAD

When a specialized expression, such as "ozone layer" or "black hole," contains words in common usage elsewhere, problems still arise. Many people visualize the ozone layer as a thin vapor in the atmosphere that is inches, feet, or perhaps yards thick. The word "layer," taken out of context, is the culprit here, since we think of layers as relatively thin regions, such as a layer of clothing or a layer of a cake. When it is used in the specialized expression "ozone layer," our intuitive understand-ing fails us.

The traditional ozone layer is the lower stratosphere. The ozone up there is in a "layer" some 12 miles thick, starting about 9 miles above the Earth's surface. The ozone is a very small fraction of all the gas in the lower stratosphere. If you were to take all the ozone from that area and compress it to the density of the air we breathe, it would make a "layer" only about 1/8" thick. It is a very minor component of the atmosphere, but utterly crucial to the existence and well-being of life. The importance of the ozone layer is that ozone prevents lethal UV radiation emitted by the Sun from reaching the ground. Indeed, the development of the ozone layer, several hundred million years ago, was essential before animal life from the oceans could migrate to the Earth's solid surface.

There is now a second ozone layer, created primarily by human-made pollution and lying directly above some parts of the Earth's surface, mostly over industrialized regions. This ozone also blocks UV radiation from space, but it has the secondary effect of causing breathing problems as well as erosion of rock and other substances on the ground. These problems occur because ozone is a very chemically reactive molecule. Contrary to the beliefs of some people, who are familiar with it from news broadcasts about ozone in the air, this lower area is not "the" ozone layer essential for the protection of life from UV. The lower, pollution-driven ozone layer was only made over the last few centuries.

Perhaps the most striking example of how common words used in specialized expressions create misconceptions is the concept of a black hole. The word "black" creates the impression that black holes are completely dark, meaning that no light or other radiation or matter can come out of them. This is not true. British astrophysicist Stephen Hawking has shown that black holes actually create light and matter out of "empty" space around them. You can read more about this in chapter 7 of Stephen's book *A Brief History of Time*. The chapter is entitled, "Black Holes Ain't So Black." The light and matter created outside black holes is not a "something for nothing" proposition. Rather, the black holes give up their mass in the process and thereby shrink over time, which is why I stated earlier that it is a misconception to believe that black holes last forever.

Staying with black holes for a moment, the word "holes" in their name creates the additional misconception that they are holes in space—empty regions. In fact, just the opposite is true: to form and be maintained, a black hole must contain a concentration of matter denser than any other matter in the universe. Its effects are to bend or distort the space it is in so much that nothing can escape directly from the region around it except gravitational force. Indeed, this matter is so densely packed that it no longer obeys the laws of physics that apply to matter in the rest of the universe. Trying to understand the nature of matter inside black holes has been part of the motivation to search for more complete laws of nature than we have presently discovered.

Besides the innumerable words used in science that have many meanings in everyday use, there are also many words that have limited meanings even to nonscientists but that most people never get correct in the first place. Common among these is "pulsar."

Consider your idea of a pulsar. The name originated in the observations, first made in 1967, that certain objects in space emit pulses of radiation, such as radio waves or visible light. Because the word gives the connotation of a pulsating object, most people visualize a pulsar as something that gets larger and smaller, thereby emitting a changing amount of energy. However, many pulsars emit pulses once a second or even faster. Calculations revealed that objects pulsing that fast—by expanding and contracting—would immediately blow themselves apart. Within a year after their discovery, Thomas Gold, an astrophysicist at Cornell, proposed that pulsars are created by the rotating remnants of very massive stars that, like the Earth and our Sun, have magnetic fields that do not emerge along the axis of rotation (figure 2.2). As the magnetic field sweeps around, it interacts with nearby gases and thereby creates a beam of electromagnetic radiation, like a lighthouse beam, that may include radio waves and visible light, among other types of radiation. If this beam sweeps across the Earth, then we see it as a pulse of electromagnetic radiation—a pulsar. (If the beam doesn't cross the Earth's path, then we don't see the pulsar.) This is now the accepted theory of pulsars.

Similar-sounding or similar-meaning words are another source of incorrect beliefs. Many people confuse "meteor," "meteorite," and "meteoroid"; "comet," "meteor," and "asteroid"; "solar system" and "galaxy," "galaxy" and "universe," "novas" and "supernovas," and "rotation" and "revolution," among others. Here are some definitions to clarify the meanings of these words.

Meteoroids are pieces of rocky and metallic interplanetary debris smaller than asteroids. *Meteors* are pieces of space debris that are vaporizing (at least in part) in the Earth's atmosphere. *Meteorites* are pieces of meteors that land intact. Most meteors completely vaporize before they reach the ground. The dust they create helps water to condense in the air to form raindrops, and eventually this debris reaches the Earth along with the precipitation. Space debris falls on your head every time you take a stroll in the rain. Keep in mind that a *falling or shooting star* is not a star falling to Earth at all. Rather, these are just alternative names for meteors.

A *galaxy* is a group of between 10 million and 10 trillion stars, along with accompanying interstellar gas and dust and other matter whose nature is not yet known, all bound together by their mutual gravitational attraction. Our Milky Way galaxy is but one of an estimated 50 billion galaxies in the visible universe.

Supernovas are not just more powerful novas. Supernovas are the explosions of stars that have more than about eight times as much mass as the Sun. The details of this process are very interesting but would take us well beyond the scope of this book. Let it suffice to say that a supernova is so powerful that for a few days the explosion can outshine billions of ordinary stars. *Novas*, on the other hand, are relatively wimpy explosions created on the surfaces of white dwarf stars orbiting a companion star. A typical *white dwarf* is an Earth-sized remnant of a star composed almost entirely of carbon. When enough hydrogen from the companion star's outer layers is pulled onto the white dwarf, this hydrogen begins fusing and, like an oversized hydrogen bomb, explodes and thereby ejects itself into space as a nova.

Finally, in the realm of how words can create confusion, consider what happens at parties or on airplanes when people ask me what I do. When I tell them I'm an astronomer, I am often asked if I'll do a read-

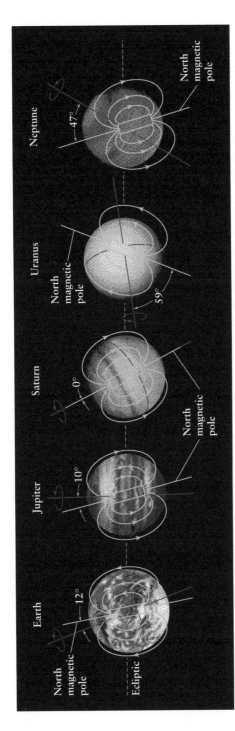

FIGURE 2.2 The magnetic fields of most of the planets do not emerge at the planets' rotation axes. The origin of this phenomenon is still under study.

ing or what my sign is (Taurus by the traditional astrologer's zodiac, Aries by the astronomical zodiac). When I say I am a cosmologist (someone who studies the overall evolution of the universe), I am often asked where my beauty shop is.

Mythical Concepts

ASTROLOGY

Consider astrology as one example of the acceptance and use of incorrect beliefs. Astrology asserts that objects in space control or influence our lives. I can conceive of two physical mechanisms by which stars, planets, and other astronomical bodies might affect us at birth and thereafter: by their gravitational force and by their radiation (meaning their radio waves, heat, visible light, ultraviolet radiation, x-rays, and gamma rays). Let's consider these possibilities.

A simple calculation reveals that the gravitational force from a person standing next to the delivery bed or the gravitational force from the building in which the delivery occurs has more effect on a newborn child than the gravitational force from any of the planets. Indeed, the effect of a nurse walking toward or away from the bed in which you were born had a greater gravitational effect on you than that of all the planets combined.

Every day of your life, everywhere you go, the mass of your physical surroundings on Earth has much, much more gravitational effect on you than do the planets. The Moon and Sun certainly create tides on Earth, but the tidal effects they generate in our bodies are absolutely overwhelmed by the forces of the atomic bonds that hold us together and the motions of the atoms in our bodies created by our internal temperature.[6] So different people born in different places, even within minutes of each other, will be influenced by the gravitational effects from the "stars" much less than by the gravitational tugs of the earthly environments surrounding their birthplaces.

---

[6] Heat manifests itself by making atoms and molecules vibrate or flow. The hotter an object, the more its atoms and molecules move.

Radiation from space is indeed something to be reckoned with, but most of what is emitted by the Moon and planets is just radiation from the Sun scattered by their surfaces or clouds, composed mostly of visible light and heat. The level of the lights in your delivery room had a much, much greater effect on you than the radiation from the Moon and all the planets combined. Indeed, even sunlight could only have affected you at birth if it had been shining on your face and eyes for enough time to cause you to go blind. Since most people are born in rooms with closed windows (or none at all), this is very unlikely. There is no known physical effect from objects in space that would give their locations at your birth or at any other time power over your destiny. The fact that the planets are invisible during the day is a good indication of how weak their radiation is by the time it reaches us.

"But what about unknown physical effects? What if there are some presently unknown properties of planets and stars that can affect us?"

Fair question. In the context of Occam's razor, let's explore whether such an as-yet-unknown physical mechanism is necessary. Do we need an extra, external force from astronomical bodies to explain anything about the cosmos, including our own behavior? Or can we comprehend nature and lead our own lives based on the activities of our minds and bodies and on our interactions with the rest of the world just under the known forces?

There is absolutely no scientific evidence whatsoever to require hypothesizing a new force in nature from the stars, planets, Moon, or Sun to explain anything in astronomy or to determine our personalities, actions, or fates. Furthermore, careful scrutiny of astrological predictions show that they are as successful as they are because they are so general. If you try to get specific information from astrology or another pseudoscience, such as the date of your death, you will get predictions no better than those chosen at random.

MAGIC

Other mythical concepts include the belief that magic exists as a supernatural force, rather than as sleight of hand (I wonder if the wonderful *Harry Potter* series of books is creating the belief in magic as a supernatural force in the minds of any young readers); the belief in ghosts; the belief in the power of witchcraft; the belief that the prayers of others can cure people;[7] and the belief in miracles.

At the risk of oversimplifying a very complicated matter, let me propose that the belief in mythical concepts begins in childhood with "magical thinking." Early on, before children have any grasp of how the natural world operates, they develop the desire for things to happen for their own benefit. Children frequently invoke magic as a means to an end that they can't otherwise achieve. Unfortunately, this kind of thinking can be reinforced if the child gets what he or she wants after an intense episode of trying to mentally force it to occur.

The belief in magic usually goes away as we grow up, but there is inevitably a residue left in our adult minds. It takes the form of "wishful thinking." At some level most of us know that wishing doesn't make things happen, but that doesn't stop anyone from trying.

Misinformation

We suspend disbelief under other circumstances, not only when we encounter science fiction and fantasy. In fact, there are many times when we go out of our way to accommodate new information, such as accepting "facts" we are told by our parents, teachers, authority figures, and trusted peers. Like saying there is no Santa, it may sound like sacrilege, but all these sources are sometimes wrong.

Believing incorrect factual information is a more complicated issue than meets the eye (or ear). At the most basic level, it prevents us from developing an accurate understanding of the natural world. Suppose,

---

[7] There certainly is scientific evidence that a positive outlook can help people during times of illness. If personal prayer contributes to that attitude, then one can reasonably conclude that prayer helped in the recovery. If you know that people pray for you, that can help your attitude too. But this is different from the possibility that you can be cured if people pray for your recovery from an illness without your knowledge. Indeed, without question, people who use prayer rather than medical attention as a cure have caused countless unnecessary deaths and much unnecessary suffering.

for example, you were told that there are two billion people on Earth and that our world can easily sustain seven billion. You might well reason that even after the population explosion of the twentieth century, there is still plenty of room and resources for more people, possibly for centuries to come. However, if you had been told that the population is over six billion (which it really is) and growing rapidly, and that the maximum sustainable number is under seven billion (used for example only—the actual number is debatable), then you would probably come to a completely different conclusion.

One of the nice things about factual information is that you can verify it, should you choose to. The problem is, when the first source is allegedly trustworthy, we are unlikely to look further. Then the conclusions we draw from the incorrect information start to seep into our perspectives on other matters. This is one of the key problems with incorrect information—it leads to incorrect beliefs. Following the population example a little further, as long as you believe that the Earth is far from being saturated with people, you are less likely to support population control, improved land management, and any number of other efforts that will be essential when the population gets uncomfortably close to the maximum that the Earth can sustain.

Suppose that after years of believing an incorrect set of numbers and building conclusions based on them, you suddenly learn more accurate numbers from an unimpeachable source. How do you respond? Do you suddenly replace all the conclusions and beliefs you developed from your previous numbers with more accurate ones? I don't think so. After all, the conclusions you have drawn over the years have helped to determine many of your attitudes, not just what you think about population control and land management but also your political affiliation, your attitude toward hunting, the strength of your religious beliefs, your career, the types of people with whom you like to associate, your goals, and more. Instead, I propose that you would take the new information and massage it to fit your prior beliefs. That is much easier than having to rethink and reshape many of the important (or even not so important) beliefs that are part of your world view.

The same thing happens when learning about the cosmos. Recall the example I gave in the first chapter of the belief that the planets have circular orbits. If you subscribe to this, the reality of learning that Neptune and Pluto exchange roles as the most distant planet from the Sun comes as a surprise. Before accepting the reality of elliptical orbits, people have to accept the fact that the solar system is less symmetric, and therefore less aesthetically pleasing, and seemingly less stable and secure, than they previously thought. The appeal of symmetric, circular orbits was one of the reasons why more than 1,500 years elapsed between Ptolemy's scheme for circular planetary motions and Johannes Kepler's demonstration, based on Tycho Brahe's observations, that planetary orbits are elliptical.

Consider as another example the common (incorrect) belief that the asteroids in the asteroid belt were once part of a planet. Suppose you believe it and I tell you that it cannot be true. We know this because the total mass of the belt is less than $1/1000$ the mass of the Earth, or less than $1/10$ the mass of our Moon. Furthermore, the disruptive gravitational force from nearby Jupiter would have made it very difficult for debris in that part of space to collect into a localized region and then fall together into a single, large body.

You might respond that perhaps another planet formed elsewhere and drifted into the region we call the asteroid belt. Perhaps it was destroyed by impact with another large body. Building on the momentum of this theory, you might argue that there is so little mass left in the asteroid belt because the impact that destroyed the planet also blew much of the mass away, into other parts of the solar system. Eventually much of the debris struck other planets and moons. Not bad. Such a model probably could be made to work, meaning that it would not violate any physical laws related to matter and orbits. The problem is that this theory violates the principle of Occam's razor. The distribution and composition of the matter in the asteroid belt is so far entirely consistent with its origin as much smaller bodies. There is no scientific *need* for there ever to have been a full-sized planet in the region of the asteroid belt.

"Yes, but it could have been there."

Errors in Textbooks

As a student, I used to feel betrayed when I read a textbook and discovered a clear-cut mistake. Thereafter, I took everything in the book with a grain of salt. The first such error I recall was a simple one in a seventh-grade social studies book. Don Quixote's jousting at windmills was used to introduce some of the features of Spain, and the pronunciation of his name was given as "Quicksut." Even that little mistake was enough to raise real doubts in my mind about the book's credibility.

In a more scientific vein, my son Josh was involved in a third-grade "space night" presentation for friends and families. It was a wonderful, busy time, full of interesting projects hanging from the school's ceiling and walls. Josh's planet was Jupiter. He had a number of accurate statements in his little essay, except for one. He had read in a book that Jupiter's Great Red Spot had a temperature of 6,000°F. On the surface, this seems plausible—the Great Red Spot is red, suggesting that it is hot. Things that glow red hot are typically around 800°F. The problem is that the Great Red Spot is not glowing red hot. Rather, it contains molecules that scatter red light from the Sun into space, just like blood scatters red light. The Spot is entirely a system of swirling gases, like a hurricane on Earth. The gases that create its red hue are not known for certain, but they are probably compounds containing phosphorous or sodium, and are much cooler than 6,000°F.

Intrigued, I borrowed a copy of Josh's third-grade science text (not the book with the Great Red Spot error) entitled *Science* (Menlo Park, Calif.: Addison-Wesley, 1989). The errors I found in its physics and astronomy sections were:

- "There are three different types of force: gravity, friction, and magnetism." In fact there are four, and possibly five, forces; the ones that affect us daily are the weak and strong nuclear forces, electromagnetism, and gravity. Despite our everyday experience, gravity is by far the weakest. Friction is the result of electromagnetic interactions between the particles in adjacent objects, and it is not considered a fundamental force, like gravity. Magnetism is only part of the electromagnetic force.

- "Unlike the first four planets [Mercury, Venus, Earth, and Mars], Jupiter is made mostly of frozen gases," and "Like Jupiter, it [Saturn] is made mostly of frozen gases." This is an oxymoron, because frozen gases are solids. Water ice is, of course, frozen molecules of water. In fact, most of Jupiter and Saturn is liquid hydrogen at temperatures between 5,000°K[8] and 20,000°K, while room temperature is about 300°K.

- "Saturn has 17 moons. One of them [Titan] is almost as big as Mercury!" Suggesting that moons can be bigger than planets is not as ridiculous as it seems on the surface. In fact, the text is wrong because Titan is bigger—it has a larger diameter—than Mercury, as is Jupiter's moon Ganymede. Indeed, Ganymede has an even larger diameter than Titan, but this is not mentioned in the text.

- A figure shows Pluto as always being farther from the Sun than Neptune. In fact, Pluto's orbit is so elongated that for about 20 years of its 249 Earth-year orbit around the Sun it is closer to the Sun than Neptune. This occurred most recently between 1979 and 1999.

- "When a comet gets close to the Sun, the frozen gases in the head melt." Again those mysterious frozen gases. In fact, when ice melts, it becomes a liquid. Comet ices don't melt, they sublimate—turn directly into gas.

- "As the comet gets close to the Sun, the gases make a tail." This is true, but the omission of a discussion about the tails always pointing away from the Sun helps propagate the belief that comet tails trail behind the comet.

---

[8] Temperature in degrees Kelvin (K) is simply 273 degrees cooler than degrees in centigrade (°C). An object at 0°K is at -273°C; equivalently, 0°C is at 273°K (or 32°F).

- A drawing shows the asteroid belt (as in *Star Wars*) with closely spaced asteroids. The book also implies that all asteroids orbit the Sun between Mars and Jupiter. Actually, some have orbits that cross the paths of planets, including the Earth, while others are in exactly the same orbit as Jupiter.
- "Some scientists think asteroids come from a planet that exploded. Others think they are from two planets that crashed together." As mentioned above, neither of these theories is considered viable.

Shortly thereafter, I was helping my son James with a calculus problem when I came across this question on page 89 of *Calculus* by Ross Finney, Franklin Demand, Bert Waits, and Daniel Kennedy (Menlo Park, Calif.: Scott Foresman Addison-Wesley, 1999):

Free Fall on Jupiter The equation for free fall at the surface of Jupiter is $s = 11.44r^2$ m with t in seconds. Assume a rock is dropped from the top of a 500-m cliff. Find the speed of the rock at t = 2 sec.

As I will discuss shortly, Jupiter doesn't have a solid surface.

LEARN BETWEEN THE LINES

So far I have implied that the incorrect "factual" information we get from others is shared in good faith—that it is given from an unbiased perspective. The person who told you that there are two billion people on Earth might have truly believed that datum and not have been trying to intentionally lead you astray. While this assumption may be true in some cases, much of the factual information we receive from authority figures certainly has been filtered to reflect their beliefs, interests, and goals.

Preprocessing of information is a bigger problem in nonscientific than scientific realms of learning. For example, you might well expect to learn about twentieth-century history differently from a professor with a left-wing political belief system than you would from one with a right-wing belief system. But the problem of bias does not vanish in science. If an astronomy teacher or author is extremely interested in the history of science, they might teach you astronomy from a historical approach, emphasizing figures such as Ptolemy, Copernicus, Kepler, Galileo, Newton, Einstein, Hubble, and others more than a teacher or author less interested in history would. The reality is that there are only a finite number of hours of instruction (or reading), so the price you have to pay to learn this historical background might be learning less about what make stars shine or black holes evaporate than if you'd been taught by someone more interested in modern mechanisms.

SHAKING UP YOUR WORLD VIEW

Imagine yourself on the first day of a new class. Suppose the teacher comes in, introduces himself or herself, outlines the goals of the class, and then announces that at least 10 percent of what you are going to learn is wrong. How would you feel?

This is precisely what I do every semester when I teach introductory astronomy. On the first day of class, with 250 students watching me expectantly, trying to judge what they are in for, I make that announcement. "Obviously, I am not going to intentionally give you false information," I add. "The problem for both of us is that astronomers' understanding of the cosmos is changing so rapidly that some of what is accepted as correct today will be shown to be incorrect. I will correct myself whenever I learn of changes in our body of knowledge." Virtually all of the corrections I've provided over the years have been such details as the numbers of moons orbiting various planets, the details of the formation of new star systems, cosmological observations of distant objects, and other things that are either on the edge of what we can observe or on the edges of our physical theories.

What I am doing by making this announcement is putting the students on notice not to take everything I have to say as gospel truth. I believe this should apply to every source of information.

News Media

A Crystal Triangle

One of the most important and complex sources of our information and misinformation about astronomy and other natural sciences are the news media. Media information is so important because it is ubiquitous. Whether from television, radio, print, or the Internet, news stories about science are often our first sources of information about interesting and important scientific discoveries. As long as we trust the news source, we are apt to accept the validity of its reportage more readily than if the same information came by word of mouth from a friend or from a novel. Friends were the first to tell me that President Kennedy had been shot, but I didn't really believe it until I heard it from Walter Cronkite, then the *CBS Evening News* anchor.

The complexity of this source of information and misinformation stems from a complicated and tense interpersonal triangle among ourselves as receivers of news, the people who report the news, and the sources of news. Scientists and the media have a true love-hate relationship. Most scientists love to have their work publicized but often hate the way the news stories come out. Professional science reporters love gaining access to sources of fresh, exciting stories, but they hate being manipulated or misguided by scientists who have hidden agendas. Scientists who provide timely, interesting, and accurate information are courted by reporters and therefore have easy access to the media. However, as is true in many other realms of life, these relationships often sour. This is likely to occur when reporters discover that their scientist-contacts are poor communicators of their work or do not provide accurate, complete, or useful information. Conversely, scientists stop providing information when they perceive that their work is presented with a biased slant or distorted in the press.

As an example of the latter, scientists often believe that important scientific results are obscured in news reports, while minor research points are given overblown coverage. Consider the Hubble Space Telescope, for example. It was designed primarily to study the cosmos far from the Earth, including distant galaxies, quasars, stars, and interstellar gas, among other things. Yet most of the results about distant objects are less interesting to the public than results about planets. Therefore, planetary data from Hubble get a much larger percentage of press coverage, relative to the amount of time observed, than do data on more distant objects. For example, did you catch the press release on the October 1999 results of Hubble's study of bulges in spiral galaxies? Or how about the incredibly significant discovery of intergalactic hydrogen? The latter is important in helping astronomers understand the distribution of matter and determine the fate of the universe.

Hubble is also excellent at producing spectacular pictures, which understandably get maximum press coverage. Yet the underlying science related to these observations rarely gets disseminated. Most people saw the wonderful images of the gas pillars in the Eagle Nebula; few nonastronomers understand the important information about star formation gleaned from them. These images create an impression of the glamour of science in the public mind that is not entirely realistic. The process of transforming most telescope data into accurate and meaningful images is long, involved, unglamorous, and exacting. Make a mistake in one of dozens of parameters or steps in the analysis and you will get inaccurate images.

Nevertheless, scientists often want to know why their important observations, experiments, and theories don't get media coverage. The fact is, most of what we do would appear boring to the general public. It would take much more air time or many more column inches to explain than the media have available. Also, much of what scientists do is more tentative than the media care to tell or the public cares to know about.

The reality is that *everything in science is tentative.* Tomorrow's experiment, observation, or theory may well show that current beliefs need revision or replacement. However, our minds work differently. Most of us usually take what we hear or see and accept it as "fact." Once we do this, it is very hard to change our belief on the subject. There was a Hubble photograph that appeared to show an extrasolar (i.e., outside of our solar system) planet being ejected from its birthplace, a binary star system (pair of stars orbiting each other).[9] The Space Telescope Science Institute, which runs the Hubble Space Telescope, reported this finding even before it was published in a refereed journal (which it eventually was). Virtually everyone who heard or read of this discovery therefore believed that such a free-flying planet had been discovered. However, further observations indicated that this object isn't a planet at all. Rather, it is probably a background star superimposed on the closer binary star system and its associated gases. A retraction was published by the Space Telescope Science Institute, but this meant that the media had to acknowledge that a previously reported piece of news was wrong. The news in the retraction was also much less exciting than the concept of an ejected planet, so it got much less press coverage than the initial discovery. Furthermore, many of the people who did learn of the correction promptly forgot it, retaining the belief that the object was an ejected planet.

Such experiences contribute to the love-hate relationship between scientists and the media. Many scientists want their work publicized, for either fame or the possibility of increased funding, but they often question the accuracy of the news items based on their work. Likewise, media people are aware that different scientists have different temperaments and have to be "handled" in different ways.

The relationship between the media and us, the public as consumers, is another complicating factor. Experience shows that news consumers respond more to what they perceive as "exciting" or even sensational information than to solid but boring scientific results. Which would you rather hear about: the possibility that a "large" asteroid will strike the Earth (that event will almost certainly happen, but probably not for tens of millions of years) or that surface scientists have developed a new way to deposit copper on a silicon substrate? In fact, the latter is much more important to your life because

---

[9] More stars are formed in binary systems than are formed in isolation, like the Sun. Indeed, at least half of all the objects you see as "stars" in the night sky are actually binary star systems. This means that two thirds of all the stars in our neighborhood of the galaxy are in binary star systems.

it would enable computers to run faster, cooler, and more cheaply, but it sounds boring.

Therefore, reporters and news writers have to find a balance between exciting material and sensationalism. A red warning light should go off in your head every time you hear of a scientific "breakthrough." These are truly rare events, since science is mostly an accumulative process rather than a continuous series of profound changes. When we news consumers hear too many sensational claims in the media, we grow wary of both the reporters and the scientists whose claims don't hold up. This is a healthy response, but it can also prevent us from appreciating valuable scientific information when it does come along.

A Minute Forty-five

Misunderstandings that develop from news sources can be analyzed at another level. Most news media operate under a variety of constraints that frame the news as it is disseminated. Perhaps the greatest constraint for television and radio is time. A typical half-hour television news broadcast contains about twenty-two minutes of information, covering about eight items. When science does make it to "prime time," the item gets less than the average amount of air time—roughly a minute and forty-five seconds. That isn't enough for the reporter to give you both sufficient background and sufficient depth to make sense out of anything. Typical television coverage of a science news item provides as much information as you read on one page of this book.

There are, of course, hour-long news shows, but even their time constraints are significant. Better yet are documentaries about just a few items, or documentaries exclusively devoted to science. Even in these best of circumstances, I am told by friends and acquaintances in the industry, a large fraction of each science news item must be omitted.

News magazines have similar issues, in the form of how many words or column inches can be devoted to any given article. Typical science news articles in *The New York Times* equal about two pages in this book.

To Inform and Entertain

So what information about science do we get from the media? To keep our attention and their ratings, the media usually provide the most entertaining and provocative information *as determined by the reporter and his or her editor.* With rare exceptions, neither of these people is a research scientist who knows the "big picture" about the scientific specialties they are asked to report on. Even the best science writers know only a few fields of science well. As a result, we are often provided with "gee whiz" information, such as Hubble Space Telescope images, that excites but is not the crucial scientific result. Indeed, we are often misled about the importance of many findings by emphasis on the wrong issues.

Related to the "gee whiz" factor is the perceived need of reporters to give a "balanced" perspective about news. In the normal scheme of things, this makes sense. Hearing other points of view often helps us understand controversial issues. With few exceptions, political issues have at least two valid perspectives, even though most of us tend to side firmly (blindly?) with one or the other.

This is often true in science, as with the question in the mid-twentieth century of whether the universe formed from an initial explosion (Big Bang model) or has existed forever (Steady State model). Predictions and subsequent observation eventually revealed that the Steady State model is incorrect. However, in science there are many theories, observations, and experiments that are much more strongly supported by one scientific interpretation than by any other. That doesn't mean this interpretation is "right," but rather that all other present explanations are considered fringe points of view by most of the scientific community. Today, for example, there is no scientifically viable model of the creation of the universe besides the Big Bang, so if news coverage is given to alternative theories of how the universe formed, they should be considered suspect. The problem is that when alternative viewpoints are presented in science news, it is often hard for the public to know how reputable they and their sources are.

Even when reporters have identified the crucial concepts scientists are studying, they often do not have the time either to digest what they have been told or to explain the concepts well enough for their listeners or viewers to comprehend the underlying science. A superficial discussion of an interesting topic will often lead the public to fill in the blanks with their own homemade explanations, which frequently contain incorrect beliefs.

It is worth noting that some publications purport to provide news but are only out to make money by exploiting human gullibility. Anyone who believes what they read in the tabloid newspapers, *Star*, *National Enquirer*, *Globe*, or *Weekly World News* is bound to develop an incorrect understanding of the natural world.

The Internet

More and more of our information is coming from the Internet. I probably go to it a dozen times a day for information and data from NASA, the Jet Propulsion Lab, the Space Telescope Science Institute, and other sources, as well as to read newspapers such as *The New York Times* and *Washington Post*. But virtually anyone can put anything on the Web.

Allegedly scientific ideas are posted on the Internet without being reviewed by competent scientists. Even in monitored "chat rooms," the technical competence of the monitor is often unknown to the participants. In contrast, whenever we scientists want to publish results of our work in respected and widely read scientific journals, the papers must be reviewed by one or more specialists first.

Today, anything goes on the Internet, pretty much as was true in parts of the American West throughout the late nineteenth century. You can buy infinitely more snake oil and other miracle cures on the Web than you ever could in Dodge City, Kansas. The problem extends far beyond unscrupulous people wanting to steal your money. Many people and organizations who have Web sites knowingly or otherwise want to steal your mind. There are untold thousands of sites containing utterly inaccurate information about science (and every other aspect of life, of course).

It is incredibly important to withhold judgment about what you read on the Web. Even sites with excellent provenances often contain incorrect information posted by nonspecialists. For example, press officers for scientific organizations are not usually Ph.D. scientists; therefore, they too can make unintentional mistakes that get posted. These errors range from emphasizing different points than scientists would to entering wrong numbers or units that subsequently explain things incorrectly.

At the many sites with unknown or questionable reputations, we don't know what agendas people posting their ideas have. These intentions often become clear, but not always. For example, some of the slicker "creation science" sites look legitimate and sound scientific but contain numerous arguments that are logically inconsistent or inconsistent with accepted science. Incorrect information from such sources can create dissonance at several levels. First, when you read information from an allegedly reliable source that is inconsistent with what you believe to be correct, it may cause you to question sound beliefs. This is, of course, just what the site owners want you to do.

Second, if you read something on a topic you knew nothing about from a source providing wrong information (intentionally or otherwise), you will have a much harder time accepting the correct information when you learn it later. This is because our minds are wired to accept as correct the first information we receive about new subjects. Therefore, it is very important for children to learn accurate information about the natural world. Their minds act as sponges for knowledge, and when initially taught incorrect information, they have much more trouble comprehending correct science (and other topics) later on.

Third, once you get burned by incorrect information from a source, you have much more difficulty accepting correct information from it. By and large this is a healthy response. However, the amount of valid online information is so great that it would be a shame if you felt you couldn't trust any of it.

The first line of defense against accepting incorrect information from the Internet or anywhere else is to maintain a healthy skepticism. The second is to be sure of your sources. The third is to check with other, reliable sources. The bottom line is to verify things for yourself.

It's nice to have a list of sources that you can blame for your own incorrect beliefs and ideas about the natural world. But once information from these sources gets into our minds, we have responsibility for processing it and accepting or rejecting it. I will explore two determinants of our individual responsibility for incorrect beliefs: sources of these ideas that are not preprocessed by other humans and ways our minds can fail us when we have information to evaluate—from any source.

3

Creating Your Own
Private Cosmos

INTERNAL AND MIXED ORIGINS
OF INCORRECT BELIEFS

I have just dropped a hammer and a pigeon feather simultaneously and from the same height. You won't be at all surprised to learn that the hammer reached the ground first. This result squares with our expectations about motion through the air. In the summer of 1971, *Apollo 15* flew to the Moon. Standing on the Moon's virtually airless surface, astronaut David R. Scott simultaneously, and from the same height, dropped a hammer from his right hand and a falcon feather from his left. They both landed at the same time. This result flies in the face of normal understanding about motion, which we develop by integrating years of experience with moving objects.

Most of the information we get from various media and from our own reasoning, and the raw data about the natural world we get from our senses, doesn't just sit in our brains as isolated facts. Rather, we evaluate the new information by using both our common sense and the facts we've accumulated about related topics. If we choose to accept the new data as correct, either because we respect its source or because our reasoning tells us to, then we start incorporating it into our understanding of the natural world.

Let's consider how a few pieces of a typical model of the solar system develop. Children are told that the Earth orbits the Sun. Trusting the source of that information rather than their own perception that the Sun orbits the Earth, they incorporate the Sun-centered belief into their fledgling world views about the Earth and other astronomical bodies. Learning that there are other planets, children typically extrapolate that they also orbit the Sun.

Now consider a new piece of information: planets have been discovered near other stars (which is true). Generalizing from the above model of the solar system, you would expect that those extrasolar planets orbit their stars and not vice versa. If I suggest that, as is likely, many of these newly discovered planets have as-yet-undiscovered companion planets, you would expect that the companions orbit their stars too. If any new planets are found at distances from their stars where water can be liquid, then we might reasonably anticipate that life exists on such worlds. In sum, we take in new information and, using what we know, fit it into our world views, thereby creating a deeper and richer understanding of the natural world.

Unfortunately, the same thing happens when you accept incorrect information as correct. You use it to build or elaborate on your model of the natural world. For example, you probably have read or seen the statement that black holes are doorways or entrances to tunnels that connect far-flung places in the universe. This is a common construct in science fiction. If you accept it as true, then you can come to several conclusions about space travel that are inconsistent with the existing laws of physics. For instance, you might conclude that when we find such black holes, we will be able to use them to travel faster than the speed of light and thereby traverse vast distances in our galaxy in hours or days, whereas such travel in regular space would take thousands of years. You might also conclude that it is possible to enter a black hole and survive. Neither of these beliefs has a basis in our understanding of the physics of black holes. But once you accept them, you are open to believing other ideas based on them: aliens have visited the Earth from faraway worlds by traveling through black holes and thereby avoiding the long travel times through normal space. If you had been skeptical about the presence of aliens on Earth because you doubted they could get here quickly, then believing in travel between black holes will help you justify believing aliens have been, or are, here.

Once we take in any information and incorporate it into our world views, we own it. Whether the data came from others, as discussed in the previous chapter, or from our own senses, it eventually becomes part of us and is our responsibility as our minds process and use it. This chapter explores how we take in information—whether preprocessed by others or fresh from the natural world—and manipulate it to come up with wrong conclusions.

Sensory Misinterpretation

YES, VIRGINIA, YOU HAVE FIVE SENSES

Contrary to what we are taught in school, the normal complement of senses is seven. Besides the traditional sight, sound, smell, taste, and touch, we can sense heat (or the lack thereof, which we call cold) and acceleration.[1] The former sense keeps us from getting burned or frozen, while the latter helps us maintain balance, feel the changing speed of a car, and thrill to a roller coaster ride.

We rely on sensory data from all seven sources continually throughout the day. As used here, "sensory data" means any information we glean directly from the natural world. It differs from most of

---

[1] It's interesting to note that in biblical times the human soul was believed to be made of seven properties that were the influence of the five planets then known, plus the Sun and Moon. Those senses were speech, hearing, smell, sight, taste, animation, and feeling. The latter two were dropped as the definition of a sense was refined.

the other external sources of incorrect beliefs because it comes to us unprocessed by other minds. Therefore, any errors that result from evaluating this information are entirely of our own making. Experiential misconceptions are those based on misinterpreting our own sensory experiences.

WHEN IS THE SUN NOT YELLOW?

Consider, for example, our perception of the Sun's color. Common beliefs are that it is yellow, white, or orange. This makes perfect sense: a quick glimpse of the Sun high in the sky (and you should never look at the Sun for more than a split second without suitable protection) gives our brains the impression that it is yellow or white. Seeing it near sunrise or sunset gives the impression that it is a distinctly orange body. So which color is the Sun? All and none of these.

The Sun emits all the colors of light, as well as the rest of the electromagnetic spectrum: radio waves, infrared radiation (which we detect as heat), ultraviolet radiation, x-rays, and gamma rays. Electromagnetic radiation travels as vibrating particles called photons. Figure 3.1 shows a model of a photon and its important properties. All photons travel at the same speed (the speed of light) and the various types of electromagnetic radiation differ only in the wavelength of vibration in the photon, as shown in the figure. Indeed, all the colors of the rainbow merely represent different wavelength visible-light photons.

However, the Sun does not emit all types of electromagnetic radiation equally. Indeed, it doesn't even emit all visible light colors equally. Figure 3.2 shows how the intensity of the Sun's electromagnetic radiation varies with wavelength. Note that the most intense colors emitted by the Sun are in the blue-green part of the spectrum. In other words, the Sun emits more photons with wavelengths that we interpret as blue-green than with any other wavelengths. You might say we orbit a turquoise star.

So why does the Sun appear yellow, white, or orange? The answer has two parts. First, most of the Sun's violet and blue light, and much of its green light, is scattered in other directions by the Earth's atmos-

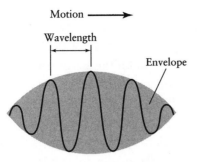

FIGURE 3.1 Schematic of a photon. Light travels in packets called photons. Each is composed of a set of waves having a fixed wavelength but varying amplitudes. All photons move at the same speed, which in empty space is the speed of light, c. The envelope is an artificial construct, which indicates that each photon is particle-like and exists in a finite volume of space.

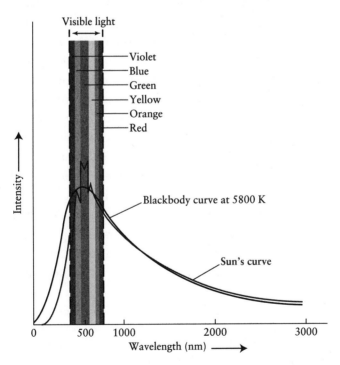

FIGURE 3.2 Blackbody radiation curve for an object at 5,800°K and blackbody radiation curve for the Sun. Blackbodies emit all wavelengths of electromagnetic radiation with a peak of intensity that depends on their temperature.

phere[2] on its way to us. This biases the information that our brains receive because we get fewer of the violet, blue, and green photons from the Sun than we would if the Earth had no atmosphere.

The second part of the answer is in our eyes, which are not uniformly sensitive to all colors. They are most sensitive to yellow and least sensitive to violet and red. Therefore, between the time that the remaining sunlight straight from the Sun strikes our eyes and the time it reaches our brains, more of the color information is lost. In other words, our senses preprocess the information they receive before it gets into the brain.

When the Sun is seen high in the sky, the most intense color that reaches our brains is yellow. When our eyes are saturated by the Sun's light, which happens very quickly, our brains often interpret its color as white. When seen low on the horizon, the Sun's light has to travel through more atmosphere before it reaches us. For example, when the Sun appears 10° above the horizon, the light is traveling through nearly six times as much air on its way to us as it would if the Sun were directly overhead. As a result, even more of all the colors is scattered by the air, including a lot more of the yellow light we normally see from the Sun when it is high in the sky. The most intense color remaining in the direct sunlight from near the horizon is orange, which is why the setting and rising Sun look orange. Therefore, the Sun is yellow, white, and orange, and it is also none of these—its most intense color is turquoise.

"It's not my fault I think the Sun looks yellow. It does. Go look at it."

Of course it does. But that doesn't make it yellow. The difference between our perception and the reality are some physical phenomena that, once understood, give us a better understanding of the Sun and other stars. In this case, the phenomena include the scatter of different colors of sunlight by the Earth's atmosphere and the nonuniform sen-

---

[2] Scattering is the process of sending light in all directions unpredictably, whereas reflection sends light on predictable paths. The atmosphere scatters violet light passing through it more efficiently than it scatters any other color. The amount of scattering for violet is followed, in decreasing order, by blue, green, yellow, orange, and red. Since violet is scattered most strongly, you might think that the sky should look violet. However, the Sun emits much less violet light than blue light, so the sky looks blue.

sitivity of our eyes to different colors. While the quick glances we have of the Sun actually do limit our perceptions of it, we would still see it as yellow when carefully studying it using neutral filters that block all wavelengths of its light equally to protect our eyes. (These filters are often used on amateur telescopes so that viewers can see sunspots.)

TWINKLE, TWINKLE, LITTLE STAR

Another astronomical example of how we misinterpret sensory information involves the twinkling of stars. We start learning about twinkling from the children's nursery rhyme. If you haven't noticed that stars appear to twinkle in real life, you might want to check it out on the next clear night. They really do seem to change intensity. If you ask an astronaut how the stars look from the space shuttle, as I have, you get a completely different story. Former astronaut George "Pinky" Nelson told me, "Their light is rock steady."[3]

The space shuttle orbits between 115 and 250 miles above the Earth's surface. If stars don't appear to twinkle at those altitudes, then there must be something between the shuttle and the Earth's surface that causes the twinkling. That something, of course, is the Earth's atmosphere.

The reason stars appear to twinkle begins with the fact that the atmosphere is not a uniform gas, even when it looks completely clear. For example, winds blow because there are regions with different air pressures in the atmosphere and the air flows from regions of higher pressure to regions of lower pressure, just like the gas in a newly opened can of soda or beer. Also, when heated by the ground, oceans, and lakes, air rises. When cooled by radiating heat into space, air sinks. Therefore, the atmosphere is a continually moving distribution of gas at different temperatures and with different densities.

Light is bent or refracted as it moves through the atmosphere in just the same way as when it goes through a lens. The constantly moving air acts as many disjointed and varying lenses that continually change the direction of the light passing through them. You have seen

[3] He also said that the Sun looks blindingly white from space.

this effect if you have ever seen the air shimmering over a hot road. Similarly, by changing the direction of starlight moving through it, the ever-moving atmosphere creates the impression in our minds that stars are fluttering or twinkling.

Another incorrect (and formerly widely held) belief derived from the misinterpretation of sensory information is that the Earth is at the center of the universe and everything orbits around it. If you or I were raised on a desert island with no book-learning or access to the outside world, I'd wager that by watching the Sun, Moon, and objects in the night sky we would come to the same conclusion.

Other incorrect beliefs that derive from misinterpreting our senses include the beliefs that the Moon gives off its own light; that distant airplanes are flying much more slowly than they actually are; that the stars and the Moon follow you in your car; that the Moon's surface is smooth; that planets change direction in their motion through the stars; that the solar system is outside the Milky Way;[4] and that the spaces between the spiral arms of galaxies like the Milky Way are empty. This last observation seems apparent from figures like 3.3, but it is so wrong that it deserves a little attention.

SPIRAL ARMS AND EMPTY SPACE

Spiral galaxies like the Milky Way have been seen all over the universe. Billions of them exist. As figure 3.3 shows, they are characterized by a central bright region called the nuclear bulge and by spiral arms. It is practically a universal belief among nonexperts that the regions between the spiral arms are nearly devoid of stars. After all, as this figure shows, those regions sure *look* empty, even when seen directly through a telescope.

I must admit that I was stunned when I first learned that there are 95 percent as many stars in the regions between the spiral arms as there are in the arms themselves. Furthermore, observations and theory have revealed that the spiral arms of a galaxy are not created by stars

---

[4] This latter assertion is explained by people who believe it by noting that the Milky Way we see in the night sky seems to be far away from us and hence outside the region where Earth is located.

FIGURE 3.3 Spiral galaxy M-83. Observe how the brightness of the spiral arms gives the impression that the region between the arms is nearly devoid of stars.

moving together in spiral patterns. Rather, the arms are ripples analogous to the waves created by throwing a rock into a quiet pond. Indeed, if you threw a rock into a rotating pond, the ripples would be spirals. We know this because back in the 1980s, astrophysicists in the then–Soviet Union didn't have the computer power available in the West to simulate galaxies. So they studied the properties of spiral galaxies by mounting pie plates on record players (remember them?), filling the pie plates with water, spinning the turntables, and dropping pebbles into the water. The ripples they saw were spiral.

In real spiral galaxies like our own, the stars, gas, and dust in the disk all travel through the spiral arms. Therefore, all the stars now between the spiral arms were once in the arms. Indeed, virtually all stars are created from gas and dust passing through the arms. Most of the new stars, moving faster than the spiral ripples in which they were created, soon move out and travel between the ripples.

If 95 percent of the stars passing through or created in spiral arms move through the arms and then through the regions between them, why are the arms so incredibly bright compared to the rest of a spiral galaxy? The answer is the last 5 percent of the stars that are created in spiral arms, but never leave them. These are the most massive and brightest stars in the galaxy. As a result of their higher mass, they create more energy in their cores and therefore shine more brightly than lower-mass stars like the Sun. Each such massive star emits over a thousand times as much light as the Sun. This light not only shines like beacons in the firmament but also causes a lot of the nearby interstellar gas and dust in the spiral arms to glow brightly. The combination of direct light from bright stars and indirect light from interstellar clouds highlights the spiral arms.

Another common, incorrect belief connected to spiral arms is that the more massive a star is, the longer it will shine. In fact, the more massive stars fuse so much matter in their cores that they use up all their fuel much more rapidly than stars with lower masses. For example, the Sun's lifetime will be roughly 10 billion years, while a star with 15 times the Sun's mass lasts only 15 million years before it explodes as a supernova. Such explosions, causing the massive stars to stop shining before they get through the spiral arms in which they were created, shed stellar mass over large regions of space. The remaining 95 percent of stars that *do* enter the interarm regions are so much dimmer that they are virtually invisible compared to the bright stars illuminating the spiral arms.

Another common misperception created by the limitations of our senses is that the "stars" we see at night are isolated objects, that each point of light is a single star. In fact, less than half of the objects we see at night are isolated stars. The rest are pairs or small clusters of stars orbiting each other. These groups are so far away that our unaided eyes cannot see the individual stars. However, there are many things that we could see more accurately than we normally do, even in the course of our everyday lives. Let's consider some.

Inaccurate or Incomplete Observations and Information

THE ROBBER WORE A METS BASEBALL CAP . . . NO, RED SOX

Studies have shown that eyewitnesses are notoriously unreliable. This is, of course, a critical issue in the legal profession. It is also relevant to understanding how we come to develop and believe erroneous ideas about the natural world. For example, most people believe that all stars are the same color, typically white, based on their experience observing the night sky. While some stars do indeed appear white, most actually do not. Check out the pattern of stars called Orion.[5] The upper left shoulder star, Betelgeuse, has a distinctly red hue, while the lower right foot star, Rigel, is blue-white. You can see the colors best by comparing stars that are close together in the sky. The more you do so, the more you will see a variety of stellar colors.

As a second astronomical example, most people believe that all the objects we see in the northern hemisphere with our naked eyes are stars. Indeed, to the casual observer, they do appear pretty much the same except for differences in brightness and, if you now look carefully, color. But looking carefully is the key, since there are some fuzzy blobs that are distinctly unstarlike. Consider Orion again. Virtually everyone who sees it would swear that each of the "stars" in it is, indeed, a star. Not true. Next time you see Orion, check out his "sword," which dangles down directly between his legs, below his belt. (I'll never understand why anyone would wear a sword between his legs, but never mind.) Even without binoculars, you can see that the middle "star" in the sword is actually a fuzzy blob. It is diffuse and indistinct, unlike the sharp points of light that other stars, such as Betelgeuse, Rigel, and those in the "belt," make. In fact, the middle "star" in the sword is a very bright region of glowing gas called the Orion Nebula, as seen in figure 3.4a.

There is another fuzzy blob we normally perceive as a star in the constellation Andromeda (figure 3.4b). This blob is actually the galaxy Andromeda, some 2 million light years[6] away (compare to the tens or

---

[5] It is likely that you thought of the word "constellation" where I wrote "pattern of stars." The word "constellation" is indeed the common term that everyone understands, but it is not the definition astronomers normally use. A constellation comprises an entire region of the sky, along with all the stars and other objects in it. The entire sky is divided into 88 unequal-area constellations. We use the pattern definition of "constellation" when there is no chance for confusion. Otherwise, the patterns are called "asterisms."

[6] A light year is the distance light can travel in one year through empty space, about 5.9 trillion miles or 9.5 trillion kilometers.

hundreds of light years' distance of most of the stars we can see with our unaided eyes).

MITIGATING CIRCUMSTANCES

Several factors participate in limiting our observational skills. First are the circumstances under which the observations are made. If you are observing the sky on a warm summer's night, more of your energy is focused on what you see in the sky, and less on keeping warm, than if you are observing in the winter when it is 10°F (-15°C). Likewise, you will see more if you have more time and less distraction. For example, you are more likely to see details of lunar craters, ejecta blankets, the phase of the Moon, and its mountains while watching the Moon alone than if you are watching the Moon with someone with whom you are planning an imminent amorous encounter.

Expectations are another factor. Suppose you are told to go out and examine the Moon every day for a week, without any further instructions. Chances are that you will make all your observations at night; that you will notice the change in the Moon's phases; and that you will observe surface features such as the craters and the dark, relatively crater-free maria (the dark gray regions of the Moon). Unless told to look for them, you are less likely to note that the directions of the points on the crescent Moon change; that the same features on the Moon are visible throughout the entire cycle of phases (the same side of the Moon always faces the Earth); that the Moon's size in the sky varies just slightly throughout the month; that despite appearances to the contrary, the Moon is not larger when it is on the horizon than when it is high up in the sky a few hours later; or that the Moon is up during daylight hours as much as it is at night. We will return to observing the Moon shortly.

This issue of expectations applies to observations on Earth as well. Different people focus on different things. For example, when looking at cars, I am interested in the make and model, which I can typically recall for a long time, but I am absolutely terrible at remembering colors.

Another influence on what we notice and remember is that we

often lack the categories and the organizational structures in our minds to relate new information to old. We see only one Moon in the sky, but a category containing one thing is hardly useful in understanding it. Compare this to the classification schemes you have developed for various down-to-Earth items. If you like cars, then you may know them by manufacturer, horsepower, styling, performance (0 to 60 mph in 8.7 seconds), features, colors, and country of origin, among other things. If you like stocks, you may know on which market they trade, their sector, their price-to-earnings ratios, their ratings, their recent annual performances, and so on. In either case, when a new specimen appears, you have ways of classifying and comparing it to items with which you are familiar. In astronomy and many other sciences, you may not have such powerful and familiar frames of reference to use in analyzing what you see or read about.

Knowing what questions to answer is a big step forward in understanding and classifying new things. Your observations of the Moon would probably provide you with completely different information and insights if you were told beforehand to determine such things as when the Moon rises and sets each day, what hours each day it is visible during the daytime, the angular distance between the Moon and the Sun, and how that angular distance is related to the Moon's phase. In other words, you will see different things and categorize them differently depending on what you are looking for.

LIQUID MEMORY

The human memory is notoriously fallible when it comes to remembering details as time goes by—ask anyone who has written a computer program but not documented it well. Furthermore, your recollection of observations you've made in the past will change. If you saw comet Hale-Bopp, would you remember which tail was on top, the blue one or the white one?

Common incorrect beliefs from incomplete, inaccurate, or earlier observations include the belief that we see the same constellations throughout the year; that all constellations (i.e., asterisms) are in the shapes of the things they were named for; that no planets are visible

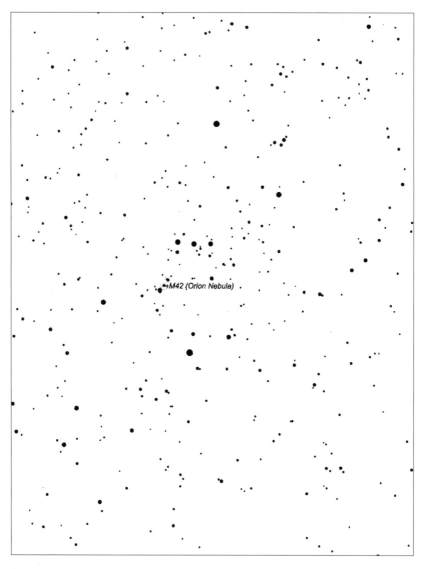

FIGURE 3.4 Nonstellar objects visible to the naked eye in the night sky: a) Orion Nebula (M-42); b) Andromeda galaxy (M-31).

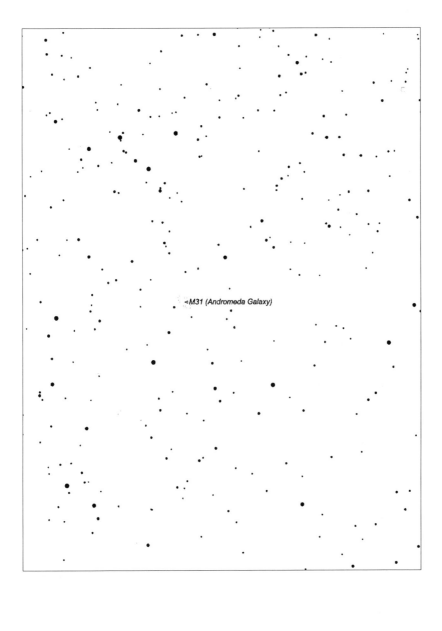
◁M31 (Andromeda Galaxy)

with the naked eye; that we see all sides of the Moon; and that the Sun always sets due west, among many others.

Our senses also have limitations and change with time. The senses in healthy young people are superbly capable of providing information necessary for survival. As we get older, our senses become less reliable, and eventually they often fail. Yet even the keenest senses possessed by a person in the spring of life are not good enough to provide reliable information about things that are very small or very distant or that do not emit signals we can pick up. Our unaided eyes cannot see viruses or bacteria, some of which are very harmful to us and other forms of life. Likewise, our eyes cannot see details of distant objects that are close together, like orbiting pairs of stars in the night sky. And even if we could catalog every visible object in our galaxy, we would be observing less than 10 percent of all the matter that must exist in it.[7] Our senses have evolved to help us survive from day to day, not necessarily to comprehend all the facets of nature. Besides the dark matter in our galaxy and throughout the universe, there are tremendous quantities of matter in interstellar gas clouds that don't emit enough visible light to be detected. Telescopes sensitive to radio waves have revealed this otherwise invisible material. For example, using the Very Large Array (VLA) radio telescope system in Socorro, New Mexico, I made images of a vast, bizarre system of intergalactic gas that fills what appears in visible light telescopes to be empty space. Indeed, this gas cloud spans the distances between many galaxies.

Besides building telescopes with which to see radiation we can't otherwise detect, we enhance our sight by developing optical telescopes that

---

[7] We know the amount of invisible, presently undetected matter that exists because of its gravitational effect on visible matter, such as stars. We can calculate the total mass that acts to keep stars, including the Sun, in orbit about the center of our galaxy. This is done by observing the orbits of stars and then using simple equations to relate the orbits to the gravitational forces that must be acting on those bodies. By subtracting the gravitational effects of all the observable matter, we can deduce how much invisible matter there is to help keep things in their observed orbits. The invisible matter is often called dark or missing mass. "Missing" is a misleading name, since it isn't missing—we just haven't found it yet. The nature of the dark matter is still under debate.

can see more detail (i.e., can separate objects that our eyes see as one), that can magnify, and that can make objects appear brighter. For example, telescopes allow us to photograph the individual stars in many binary star systems. They can also reveal billions of other galaxies that are too dim to be seen with the naked eye. Indeed, technological enhancements now exist for all our senses. These devices enable us to better understand the natural world, as well as to enjoy it and work with it.

When we possess incomplete information, due to the limitations of our senses or other factors, we don't have all the data we need to make a valid analysis and draw valid conclusions about what we sense or are told. That, of course, has never stopped anyone from convincing themselves they understand things about which they actually have less than complete knowledge.

One of the more interesting aspects of human behavior is that we tend to fill in gaps in information with ideas based on our own experiences, beliefs, and thought processes. I remember hearing about a talk by a college administrator who underscored the points he made with an impressive array of facts and statistics. After the lecture was over, a friend of his commented on how well he had supported his case. The friend went on to ask where these numbers could be found for future reference. "They can't," the administrator replied casually. "I made them up as I went along."

We go about filling in the gaps in our knowledge and our memories through a number of memory activities such as wishful thinking and common sense. While the former is fairly self-evident, the latter is incredibly complex, and its effects are far reaching. Furthermore, more often than not, conclusions about the natural world reached via common sense are wrong.

The Shortcomings of Common Sense[8]

By the time we become adults, we each develop a personal set of built-in mental scripts for quickly assessing information and drawing con-

[8] With acknowledgment to Alan Cromer and apologies to Thomas Paine.

clusions. These save time and effort, and often they are good common sense[9] and help us survive. Common sense is often defined as "sound practical judgment that is independent of specialized knowledge or training."[10] If you see a person with a gun, your script tells you to protect your loved ones at the scene and to leave the area as quickly as possible. If you are playing golf and a thunder-and-lightning storm begins, your script tells you to head for cover (but not under trees!). If you are about to drive to another city and your car's gas gauge reads nearly empty, your script tells you to get gas before you leave.

Now consider one of my favorite pitchers, Nolan Ryan, whose fast ball was clocked at 100.9 mph on August 20, 1974. Common sense tells most people that a pitched baseball takes longer to land than one just released to fall straight down from the same point at the same time. They would be wrong—both balls reach the ground at exactly the same time. Common sense tells most people that if two planets, one with half the mass of the other, were orbiting the Sun at the same distance, they would be moving at different speeds in order to stay in the same orbit. Not so. As we saw in the preface, the planets would have identical speeds.

While common sense certainly helps us in everyday life, it fails miserably in many instances and for many reasons when it comes to science. The problem with using common sense in dealing with scientific matters is that in science we encounter many things for which our reasoning is faulty and to which our scripts don't apply. Nevertheless, common sense is our first response to new situations.

For example, I frequently get phone calls from people who have seen strange lights in the sky the previous night. While some people explain their experience by saying that they have seen "bizarre lights," most say that they saw a UFO. This expression can mean a variety of

---

[9] Other scripts some of us use are neither common sensical nor helpful for survival. For example, stereotyping reduces our need to communicate and explore what other people are really like. More often than not, it leads to unjustified conclusions about people and to behaviors on our part toward them that are often inappropriate and counterproductive.

[10] *Webster's College Dictionary* (New York: Random House, 1991).

things. At face value, it means something that is flying and that the observer cannot identify. However, to many people it also carries the connotation of a spacecraft from another world. Many people get to the latter definition because they see the lights and immediately try to classify them. Common sense first says that they are from a plane or helicopter. If the motion of the lights is inconsistent with such aircraft, people often try the next obvious explanation, such as an experimental aircraft. When calls to local military air bases reveal that there were no recent test flights, the list of common-sense explanations dwindles to one in our science fiction–drenched society: aliens.

This process has several shortcomings. First, as discussed above, observations, especially of things we are not familiar with, are notoriously difficult to make and interpret. We try continuously to "place" what we see, and this colors our perceptions. Second, most people don't actually succeed in considering all the possible classifications they already know. Common sense also fails us in areas where we have less than complete understanding, such as the science of the natural world. For example, most people don't know about a variety of atmospheric and astronomical phenomena that cause changing colors, hovering lights, rapidly moving and rapidly accelerating lights, and streaks of light across the night sky, sometimes accompanied by sound, that abruptly disappear or explode. Without this information, and driven by knowledge of innumerable stories of aliens, many people leap to the UFO-alien connection.

As a more down-to-Earth example of where common sense fails, consider how we respond to a traumatic event like the shootings in Colorado at Columbine High School. I think it is fair to assume that everyone who heard of the event immediately tried to imagine what kind of people would do such things, even before the two killers were profiled in the media. Furthermore, I'm willing to go out on a limb and bet that like me, most people created inaccurate initial images of the two youths. In this situation, common sense suggests such killers would be extremists or bigots or people with a history of violent behavior. The motive of revenge against classmates is so far from the norm that it didn't fit into a common-sense script for such behavior. Such events have occurred often enough now, however, that such behavior *does* come to mind when we hear of a school shooting.

When we process new information, we bring to bear a variety of elements, each of which affects our perceptions. The range of issues we include is so complicated that I would be surprised if anyone yet knows all of them. I'll now explore several common, fundamental mistakes of reasoning that affect how many people perceive the natural world and that often lead to unjustified conclusions and incorrect beliefs.

OVERGENERALIZATION

One of the devices our brains often use to assess new objects or events is to relate them to apparently similar objects or events that we have experienced. This often takes the form of unjustified, but common-sense, generalizations. For example, if you have ever traveled abroad, you have probably said, "Hey, this place is just like. . . ." Superficial similarities often do exist between different locations on Earth. But below the surface, the differences are invariably much more numerous than the similarities.

Consider what happens the first time a person learns that there are at least sixty-five moons in our solar system. Asked to describe them without being told anything more, he or she will invariably depict them as similar to our Moon: spherical, airless, with light and dark gray regions and craters. In fact, fewer than twenty-five of the moons are even spherical. The others look more like giant potatoes than anything else. Furthermore, several, including Titan (one of Saturn's moons) and Io (one of Jupiter's moons), have measurable atmospheres. Indeed, Titan's atmosphere is ten times denser than the air we breathe. Most moons have either a single hue of gray, like Mars's Phobos and Deimos, or a variety of colors and surface features, like Io and Neptune's Triton. All the moons that have been studied *do* have craters, but we can overgeneralize about those as well. While our Moon's craters were all created by impacts, some moon craters, especially on Io, were unquestionably created by volcanos. Indeed, some of these volcanos are still active, and we have seen them erupt. It's fair to say that if I were to show you photos of all the moons, you would be able to pick out ours in a moment—none of the others looks even remotely like it (see figure 3.5).

As another example, when learning about the planets, many people assume that the eight other planets have characteristics similar to the Earth's. Specifically, common sense tells them that all the planets have solid surfaces. In fact, Jupiter, Saturn, Uranus, and Neptune all have Earthlike cores, each containing more mass than the entire Earth. However, around these solid cores are layers of liquid tens of thousands of miles thick, surrounded by thick atmospheres. If you were to land on any of these planets, you would descend through an ever-thickening atmosphere until you were crushed by the pressure of the gases around you. Then your remains would flop into the liquid. There are no solid surfaces protruding anywhere on any of these planets.

I took the liberty of writing the word "liquid" in the preceding paragraph to make a point. If you have not studied planets at all, that descriptor creates the (overgeneralized) impression of massive oceans of liquid water. Indeed, the terrestrial cores of Uranus and Neptune are surrounded with layers of water estimated to be more than 6,000 miles thick. However, above these are layers of liquid hydrogen and helium more than 4,000 miles thick. Jupiter and Saturn appear to lack the water layer completely and have only hydrogen and helium shells more than 20,000 miles thick extending from the solid core to the atmosphere.

So far, the overgeneralized concepts have pertained to concrete objects, such as that all moons are spherical (or round) or that oceans on other planets are made of water. However, often what we overgeneralize is a principle, based on our experience. Consider why so many people believe that the seasons are caused by changing distance between the Earth and the Sun. Since changing seasons imply changing temperatures, we naturally refer to our experience with hot objects to understand the varying seasonal temperatures—the closer we are to a fire, the more heat we feel. Applying this notion to the Earth–Sun system creates the common misconception about the cause of the seasons.

This overgeneralization and other misconceptions may result in some cases from even deeper functioning of our brains. Andrea diSessa, a science education specialist at MIT, has proposed that we each

FIGURE 3.5 Various moons in the solar system. Can you identify which is our Moon?

develop a set of fundamental beliefs that we can call upon when we need to evaluate new situations. He calls these basic mental resources phenomenological primitives, or p-prims. For example, knowing that the closer you are to a fire the hotter you feel comes from a p-prim that tells us in general that "closer means stronger." In this case, of course, closer to the fire means hotter. This p-prim could also be drawn upon to tell us that the closer you get to a loudspeaker, the louder the sound, or the closer you get to a light, the brighter it appears. Another p-prim, "dying away," is our understanding that without external forces acting, things die away with time. Examples include a fire going out, an object slowing down, a bell ringing more quietly.

The reality in science is that the conclusions we reach about many things using p-prims or other thinking processes aren't correct. Besides the seasons misconception that can be derived from the "closer is stronger" p-prim, an incorrect understanding of the motion of rockets moving in space can follow from another p-prim that essentially tells us that keeping something moving requires continually applying force to it. This certainly is true for most situations on Earth: slide a penny on a table and it will slow down and stop; get your car going at a constant speed and take your foot off the accelerator, and the car will slow down and stop. Using such reasoning, many people conclude that a rocket engine must keep firing for the rocket to move through space. This is a misconception. In space, there is virtually no resistance to motion, so once something gets up to speed, you can turn off the propulsion force and it will simply travel in a straight line at that speed unless acted on by an outside force, such as the gravitational attraction of a planet or other body.

Consider now an example of how poor communication about an exciting scientific discovery can create incorrect beliefs. Press releases concerning the discovery of planets orbiting other stars often leave room for overgeneralization. Many quickly describe the planet's discovery, the star it orbits, its distance from Earth, and the length of the planet's year. However, this leaves the reader to fill in at least one blank, namely what the planet is like. If you don't read down to the tenth paragraph (at least in one such release I found) you might be inclined to conclude that the new planet is similar to the Earth. So far, at least, they aren't. The planets discovered around Sunlike stars are all roughly the mass of Jupiter. Furthermore, most of them are much closer to or much farther from their star than Earth is from the Sun. If the authors of the press releases were more sensitive to the expectations and responses of their readers, this information would always be put in the first paragraph to help avoid creating incorrect beliefs.

Other examples of overgeneralization include that all planets have atmospheres; that all planets have the same atmosphere as the Earth; that all planets have one moon (this belief also comes from other sources, as we will see shortly); that all planets have the same chemical composition as Earth; that all planets have twenty-four-hour days; that all planets spin or rotate in the same direction as the Earth; that the Sun has a solid or molten liquid surface; and that all galaxies are spiral like the Milky Way.

Talk about serendipity. On the evening I was trying to think of an example to show how this behavior (overgeneralization) develops early in our lives, I was watching television with my son Josh, then age eight. A Toys "R" Us commercial came on showing a little girl, perhaps two and a half years old, announcing to everyone that she had made a poop in the potty all by herself and that, as a reward, she could buy any toy in the store. For those of you who aren't parents, trust me: making a poop in the potty all by yourself is a very, very big thing for a young child. The girl selects a stuffed animal and accompanies her mother to the checkout counter. The lines at the checkout motivate her to say, effectively, "Gee, a lot of people must have made a poop in the potty all by themselves."

While scatological, this mostly charming commercial demonstrates truly how children overgeneralize. In this case, the girl concluded that everyone buying something was receiving a reward for doing the same thing she had done. So accurately were the child's thought process and behavior portrayed that I would not be in the least bit surprised if I heard the same phrase uttered in real life.

UNIQUENESS

While overgeneralization is a common cause of incorrect beliefs, the opposite activity also leads us astray. People assume incorrectly that many things are unique. I suspect that this element of common sense stems from our own egos. That is, we each understand that we are unique individuals and, at least in most cases, we make decisions and behave according to our personal histories.

Suppose, instead of telling someone new to astronomy that there are at least sixty-five moons in the solar system, I asked them how many moons there are. There is an excellent chance that they will answer either one or nine. That ours is the only moon is one of the most common incorrect beliefs about moons, and it clearly stems from the fact that throughout our lives we see only one in the sky. A person who initially holds that ours is the only moon can equally well believe that all moons are spherical when they learn that there is more than one. Believing that there are nine moons is an overgeneralization of the Earth-Moon system to all the planets in the solar system. I mention this here because it is important to stress that different people have different priorities when they evaluate new situations.

Other common examples of the misuse of uniqueness are the beliefs that Earth is the only planet with water or with an atmosphere; that the Sun is a unique object (it is actually an ordinary star among hundreds of billions in our galaxy); that the Moon is the only object that goes through a cycle of phases as seen from Earth (all the planets actually go through cycles of phases as seen from here); that our Moon is the only body with craters (Mercury, Venus, Earth, Mars, and all the moons in the solar system have them too); that Earth is the only planet with seasons (all planets have them to some degree); that Earth is the only planet with volcanos (Venus and Mars have them too); and that the Milky Way is the only galaxy.

PERMANENCE

Many people develop the expectation that important things don't change. This begins in childhood as most children hope, and therefore expect, that they and their parents will live forever. Most adults in the developed world expect political structures, such as countries and political parties, and certain interpersonal relationships to last. This expectation of permanence is never satisfied, of course, but it is a comforting illusion that allows us to rely on these structures and relationships in daily life.

The belief in the permanence of things is frequently overgeneralized to the expectation that astronomical bodies, such as comets, the Sun, other stars, Earth, and the Moon will last forever. Also common is the belief that the day will always be twenty-four hours long, which we saw earlier is incorrect. Most people believe not only that the Sun and existing stars will always be, but also that these objects will always have the same sizes and colors they do today. Similarly, most people believe that the constellation patterns (asterisms) will always remain as they are now. None of these expectations is correct. I'll briefly consider each.

Comets.

It is likely that after reading the discussion of comets in the first chapter you have already figured out why they do not last forever. To summarize: comet comas and tails are created from gases and dust that were part of the comet. This material spreads into space and never comes back. Therefore, comets get smaller and less massive each time they orbit near the Sun. One interesting side effect of comets' disintegration is that they leave pebble- and larger-sized debris in orbit around the Sun. When the Earth passes through it, some of that matter falls to the surface, creating the meteor showers that we enjoy at various times throughout the year.

Earth, Moon, and Sun.

The fates of the Earth, Moon, and Sun are all connected. At the crux is sunlight. We saw in the first chapter that this radiation is created in the Sun's core by the conversion of matter into energy. The amount of energy created is related to the amount of solar mass lost as given by Einstein's equation: $E = mc^2$. About 5 billion years from now, all the hydrogen in the Sun's core will have been converted into helium and so fusion will cease there. But the energy generated by fusion is also what keeps the core from collapsing under the weight of the Sun's mass. Some of the energy created in the core today goes to counterbalance the inward force of gravity by pushing the outer layers of the Sun outward. Common sense might seem to imply that if the fusion in the core stopped, the entire Sun would collapse. That is not quite correct.

Without fusion, the Sun's core *will* begin to collapse, as will the gases in a shell just outside the core. As this shell of gas moves inward, it will be compressed and heated until it grows hot enough for hydrogen fusion to occur in it. Similarly, the core will become hot enough to fuse its helium into carbon. Since the fusing shell of hydrogen will be closer to the Sun's surface than the core, the energy created in it will have less matter to push outward than the energy created in the core. The result will be that the energy from the shell can push harder on the outer layers and actually force them to move outward even farther than they are today.

Therefore, about 5 billion years from now, the outer layers of the Sun will expand outward. As they do, they will get farther and farther from the energy source in the shell of fusing gas. As a result, the Sun's outer layers will cool. The Sun will turn red, since objects glowing red hot are cooler than objects glowing yellow (or turquoise). This evolutionary change in the Sun's size and color is typical of all stars—as they change size, they change color.

Calculations predict that eventually the Sun will become so large that its outer layer will extend out to the Earth's orbit and beyond. At that time, its heat will cause the Earth to lose its atmosphere and cause its surface to char. Friction between the Sun's outward-moving gases and the Earth may be great enough to slow the Earth and Moon down, causing them to spiral into the Sun and vaporize.

Eventually the Sun's core will be converted into carbon. The Sun does not have enough gravitational force to compress and heat its carbon sufficiently to start fusion. But the shells surrounding the core will continue fusing and pushing the outer layers farther and farther outward until they drift away forever. This relatively gentle shedding of mass is called a planetary nebula. The remaining carbon core, called a white dwarf, will be about the size of the Earth but much denser.

The vast majority of stars in the universe will end their "lives" this way. Those stars with more than eight times the Sun's mass evolve beyond the carbon core. They continue fusing, thereby creating elements as heavy as and including iron in their cores, and then explode as supernovas.

Constellations.

The stars that make up the constellation patterns are all in our Milky Way galaxy, but contrary to appearances, they are not all at the same distance from the Earth. While the nearest stars are just over 4 light years away, some of the bright stars, like Deneb in the constellation Cygnus, are more than 1,500 light years away. The stars and all other objects in our galaxy orbit its center.

Contrary to common belief, our star, the Sun, is not fixed in space. Rather, it is moving in a circular orbit around the center of the galaxy at the rate of about 250 million years per orbit. However, not all stars orbit at the same speed that our Sun does. Some are moving around the galaxy faster, others more slowly. Stars are found at different distances from the Earth; even stars in the same constellation move at different speeds across our sky. Therefore, over time, stars appear to drift apart or together depending on our orientation to them, so the appearances of the constellations (patterns) change in our sky.

Finding Patterns Even When They Don't Exist

We humans are remarkably good at finding patterns and associating them with familiar objects. Looking at the clouds just now, I saw a seal, a duck, a submarine, and a sea horse. Looking at the night sky, it's easy to see a human shape in Orion, a pot in Ursa Major (part of which is the Big Dipper), a bird in Aquila (the eagle), a scorpion in Scorpius, and a lion in Leo, among other familiar objects in the constellations.[11]

Part of our ability to recognize patterns is hardwired in our brains, which "automatically" identify straight lines, circles, and other basic patterns. More complicated patterns are learned and stored for rapid use throughout our lives. This is a very useful evolutionary feature. Imagine what life would be like if every time you saw a car, you had to

---

[11] It is well worth noting that most of the stars in various constellations do not make patterns that look anything like their namesakes. I defy you to see a lion in Leo Minor, a king in Cepheus, a harp in Lyra, a bear hunter in Bootes, or a little dog in Canis Minor.

reconstruct what it is in your mind, rather than taking the general pattern and almost instantly knowing you had to get out of its way.

When we deal with visual patterns, real or imagined, our brains provide possibilities that our senses help to confirm or refute. For example, suppose you saw a V-shaped pattern moving across the sky. Before it was close enough for you to identify, you would probably be holding two possible explanations on tap: a plane or a flock of birds flying in wedge formation. Confirming one or the other would free up that part of your mind to deal with other things; unless distracted by more pressing matters, you would probably wait to find out.

Besides visual patterns, we all develop expectations about events located close together in time and space. Experience with lightning and thunder teaches us to expect the latter shortly after we see the former. Such occurrences are a form of temporal pattern recognition—associating events with each other in time. When thunder and lightning, the screech of tires and the thud of a crash, or the sound of breaking glass and swearing occur, we are usually justified in assigning a cause-and-effect relationship to the two events. However, both in everyday life and in science, we must be careful in making such connections between two things that occur near each other in time and space.

True cause-and-effect relationships do occur in astronomy, such as meteor showers during Earth's passage through the debris left by a comet or the formation of a crater when debris from a meteor (a meteorite) strikes the Earth's surface. However, there are many instances in everyday life, as well as in science, when two totally unrelated events occur in proximity and are incorrectly associated as a cause-and-effect pair. For example, suppose you hear a crash outside and go out to find someone lying in the road and a car crashed into a nearby tree. You would probably conclude that the driver lost control, hit the pedestrian, and then hit the tree. In this scenario the driver caused the injury and the accident. But it easily could have been otherwise. The pedestrian could have suffered a heart attack in an unrelated incident and a drunk driver could have hit a tree, thereby having an accident.

Choosing the First/Simplest Explanation

When we are asked a question, most of us latch on to the first plausible response that comes to mind. If we see something like the accident scene just described, most of us reach the "obvious"—and the quickest, most convenient—conclusion that the car caused the injury. No doubt you have had encounters in which you were not satisfied with something you said off the top of your head and afterward thought, "I should have said *that*" or planned what to say next time.

When trying to explain something in nature that we encounter for the first time, we also tend to accept the first plausible explanation that comes to mind. If the belief we develop is even superficially consistent with what we observe, we often don't bother evaluating it further, even when we have the leisure to do so. Such first answers are also often the simplest ones, because as soon as we have applied just enough information from our memories and our reasoning to draw a conclusion, we tend to choose it and consider the subject closed.

This origin of incorrect beliefs is often connected with a variety of other origins. We use overgeneralization, object permanence, incomplete reasoning, or uniqueness to come to a conclusion that "feels" right. Choosing the first explanation that comes to mind leads to one of the shortcomings of human nature: thinking that we have a suitable explanation and failing to probe its implications. Often, inconsistencies with our first explanation arise even after just a little probing, as I tried to show for several common first explanations in the first chapter.

Classic among the incorrect first-to-come-to-mind beliefs that we develop is the cause of the seasons, based on the plausible but actually irrelevant idea that getting closer to a "fire" means getting hotter. Here are some others: that stars shine by burning something; that planets, moons, and space debris are all composed of the same elements; that planets have circular orbits; that planets move in constant-speed orbits around the Sun; that the sky is blue because it contains a blue gas; that the sky is blue from reflected water; that the Sun is hottest on its surface; and that the only purpose of a telescope is to magnify things.

Consider this example, in which our first answer combines with expected cause and effect to give us the wrong explanation. When the Sun has many sunspots, astronomers say that it is active. This is a time of maximum outflow of particles from the Sun, sometimes in the form of very powerful flares and other mass ejections, and of frequent auroras on Earth. The year 2000 was a period of maximum sunspot activity. These peaks or cycles occur roughly every eleven years. During such times, low-orbit satellites such as the International Space Station tend to move Earthward, often requiring that they be forced back up to their desired orbits.

Now, it seems plausible that the extra particles from the Sun are pushing on these satellites, causing them to move closer to Earth. I think it's fair to say that this is the line of reasoning that first comes to mind for most people. It also establishes a direct cause-and-effect relationship between the Sun and the movement of the satellites. But this explanation is wrong. The real reason these satellites lose altitude is that the active Sun's particles and radiation heat the Earth's atmosphere more than usual, causing it to expand outward (heated gases expand). The expanded atmosphere rubs against the satellites, and this friction causes them to lose energy (forward momentum) and therefore spiral downward.

Pulling up beliefs and ideas from our experience and applying them to new situations is one thing. Using correct and complete reasoning is something else, something that rarely happens.

Incorrect and Incomplete Reasoning

Our powers of reasoning are constantly being exercised. First we identify a goal, either from an external source, such as when your boss tells you to do something, or internally, such as when you decide what to wear to work. Then we draw upon our experience, training, input from others, logic, and emotions to decide how to respond. When deciding what to wear to work, we take into account such factors as the weather, the people we expect to see that day, current styles, dress codes, comfort, and sex appeal, among other things. Next we evaluate our options, namely our wardrobe, and then we act.

Such complex reasoning is a quintessentially human activity. While we are driven by biological needs and instincts, we are uniquely capable of controlling those drives and acting for other reasons. The human brain has evolved into a stunningly complex instrument capable of taking in information, storing it, recalling it, modifying it, evaluating it, learning new ways of thinking, and acting on these thoughts through our bodies.

Each element in the process of reasoning is problematic. As I've already discussed, what we take in from the outside is affected by our senses. It is also affected by how it is stored in our minds, a hot area of neurological research. As a result of the complex way in which memories are stored, most of us have less than perfect recall. What we remember is rarely exactly what we experienced, so we unintentionally fill in the gaps with "standard" images. If you are asked what clothes you were wearing on a certain date but don't remember, you will probably come up with an answer based on what you were likely to have been wearing. If you are asked the color of a car you saw several weeks ago but you only remember it was light colored, you will probably reply something like, "I'm not sure, but I think it was. . . ."

Learning new ways to think or behave is extremely difficult. For example, learning to reason using the laws of physics takes years and years of study, which unfortunately limits the number of people willing to put in the effort. On the other end of the spectrum (perhaps), learning to be a good soldier also takes years of effort and often painful reprogramming of the way one thinks. This is why military training, especially boot camp, is such a harrowing experience. The goal of military instructors is to subjugate the soldiers' free will to the often life-threatening will of the military apparatus. Teaching "stay and fight" discipline in situations where evolution and normal reasoning say, "get the hell out of here" is very difficult.

INCORRECT LOGIC

Learning to use correct logic is also very difficult, often because the rules of logic are inconsistent with the way we normally think. Most of us violate the rules of logic in our reasoning all the time. For example, just because science has not disproven the existence of aliens on Earth (which is impossible to do, by the way), that doesn't mean they are here.

"But it also doesn't mean they aren't here."

I agree. They might be. But Occam's razor dictates we do not believe in them until there is some reason to do so. In general, whatever has not been proven false *need not* be true and whatever has not been proven true need not be false.

There is also a difference between logical impossibility and physical impossibility. It is both logically and physically impossible to be a little pregnant. Either you are or you aren't. It is logically possible for me to jump from New York to Melbourne, Australia, but it is physically impossible for me to do that.

Learning correct logical thinking is a heady experience. For weeks after taking a college course in logic, I thought differently than normal. I actually felt as though I could "see" flaws in reasoning that had not been apparent to me before. However, as with most other things in life, if you don't use it, you lose it. As time went on and I became less conscientious about thinking logically, many of the rules and techniques slipped away. While some of them remain with me, many of them have been replaced with the less logically correct thought patterns I had used before. Seeing where logic breaks down is a powerful tool in altering incorrect beliefs, but it doesn't come easily. I have included some references on the subject in the bibliography.

INCOMPLETE REASONING

We humans don't normally have all the tools to reason correctly. That is, we lack perfect powers of observation, perfect memories, perfect recall, perfect logic. Compounding this is the fact that we rarely reason completely. As noted above, we tend to grab the first explanation that comes to mind without thinking through its implications and validity. When faced with new information about or experiences in the natural world, most people don't analyze all the relevant scientific facts. As a result, when evaluating a subject, they draw incorrect conclusions that lead to incorrect beliefs.

For example, introductory college physics students are taught (correctly) that the force of friction always acts opposite the direction of motion. If this were not true and the force of friction could act *in* the direction of motion, then you could use the force of friction to help speed things up. But friction always and only takes energy out of a system (as heat and sound); therefore, it cannot contribute to motion—it must act to oppose motion. So when I draw a picture of a car and ask students to show what direction the force of friction is acting on the wheels, most of them draw the force acting backward, opposite the car's direction of motion. They are wrong. The force of friction points in the direction of the car's motion.

"Now wait a minute. You just said friction always acts opposite the direction of motion."

The conclusion that the friction points opposite the car's direction of motion comes from an incomplete analysis of the problem. Stand outside the passenger side of a car and watch it move. To move a car to the right, from your point of view, its tires must be rotating clockwise. Therefore, where the tires make contact with the ground, the tires are moving to the left. To keep the tires from spinning, the force of friction between them and the ground must be in the opposite direction of *their* motion. Since the tires are moving to the left at the ground, the force of friction must be acting to the right, which is in the same direction as the car's subsequent motion. So the force of friction is opposite the direction of motion of the tires, which create the friction, not opposite the direction of motion of the car to which they are attached.

Likewise, untrained intuition is often incorrect. Many people think that a ball thrown in the air stops accelerating at the top of its motion or that a pound of lead weighs more than a pound of feathers or that heavier objects fall faster than lighter ones or that electrons stop flowing (are used up) after they travel through an electrical appliance. These beliefs are due to a lack of understanding of the laws of physics. The laws of logic and the laws of physics are two different fields of study, although an accurate understanding of the natural world ideally implies an appreciation for both.

The falling objects issue is interesting because it shows how experience can support incorrect intuition and lead to incorrect beliefs through incomplete reasoning. Recall the hammer and feather experiments that opened this chapter. In a third experiment, when you drop them in a vacuum chamber on Earth, they reach the ground together, just as they do outdoors on the Moon. Clearly, the physical effect that slows the feather under normal conditions on Earth is air friction. As objects move through the air, they rub against the air molecules, creating friction, which always acts against motion and therefore slows the objects down. To fully understand an event in the world around us, we have to understand each component of the event. To understand the motion of falling objects, we have to be conscious of the fact that gravity and air friction are two separate phenomena and of how each of them affects the event. With this awareness, I could predict that without air friction, I could drop the Empire State Building and a feather from the same height and they would both reach the ground at the same instant.

Suppose I asked you to consider two stars, one with twice the diameter of the other. Would you be justified in concluding that the bigger one has more matter in it (is more massive) than the smaller one? The answer is given on the bottom of page 118.

ANIMATING, ANTHROPOMORPHIZING, AND WHY

Buried deep in our psyches is the need to identify causes for things that happen. Put another way, we are searching for the answers to the question, "Why?" This behavior includes animating or giving living characteristics to inanimate objects,[12] and anthropomorphizing or giving human characteristics to inanimate objects or to animals driven by instinct and biological urges. Both of these actions lead to incorrect beliefs about the natural world.

Scientists are not exempt from this very nonscientific behavior, at least in the words we use for certain activities. For example, you may have noticed earlier in this chapter that I put the word "lives," as related to stars, in quotes. This was to highlight the use of words appropriate for living things to describe inanimate objects. In anthropomorphizing's most extreme manifestation, people have given objects like the Sun supernatural powers and then proceeded to worship them. Indeed, virtually every civilization has had a Sun god at one time or another. Symbolism related to these early beliefs can still be found today in most religions.

Both of these behaviors, anthropomorphizing and animating, originate in our childhoods. You can watch children "pretending" that their dolls, stuffed animals, and toys are alive (animating) and they attribute human feelings to their toys and pets (anthropomorphizing). When my own children were young, they did this with gusto. Such behavior is definitely cute and I wouldn't dream of suggesting children be taught otherwise. However, to a greater or lesser extent, this behavior carries over into adulthood. For example, many people develop an emotional attachment to their cars that goes as far as giving them names and mourning when they break down, are involved in an accident, or sold. Or when a computer breaks down, a person may feel that the action was

You are mistaken if you concluded that the larger star necessarily has more mass. In all likelihood, you came to this conclusion by assuming that all stars have the same density, an example of incomplete reasoning. Density is the amount of matter packed into a volume. To see that this isn't so, recall that I explained earlier in this chapter that the Sun will expand greatly as it ages. At the time it engulfs the Earth, it will be 214 times larger across than it is today and will take up nearly 10 million times as much space. Having transformed some of its mass into energy to continue shining, however, it will actually have slightly *less* mass than today. So even though the Sun will be much larger, its mass will be less and its density will be much lower. We would get exactly the same result, of course, if we compared the Sun as it is today with any star that was originally identical to it but has already expanded (such a star is called a giant).

So a larger star need not be more massive than a smaller one. Indeed, there are remnants of stars, called neutron stars, that are only a few kilometers across but contain more mass than the Sun, whose diameter is 1.4 million kilometers. As you can well imagine, the density of matter in neutron stars is much greater than the density in the Sun. Size isn't everything.

[12] I am not going to define what is alive and what is not. Defining life is so difficult that there is no universally accepted definition. Also, for our purposes in this discussion, your "common-sense" view of life will serve.

intentional and that the machine was "out to get me" or "chose the worst time to fail"; they may even think, "Why do these things happen to me?" It may be true that someone is out to get you and that they are using your computer to make you suffer, but the machine itself doesn't even "know" you exist, much less "want" to get you.

We also tend to inadvertently anthropomorphize and animate things just with the words we use. Changes in inanimate objects are often expressed in terms of lives or life cycles. In astronomy these include the lives and deaths of stars, the life cycles of interstellar gas, and the life of a comet or any of a hundred other things that change.

If we assume something can reason or emote as we do, then we automatically develop further expectations of its behavior based on what we expect from ourselves or other people. This is a problem especially when we develop expectations that animals have a sense of right and wrong or an aesthetic value. For example, where we see a beautiful doe grazing in a meadow, a mountain lion eyeing the same creature sees lunch.

Anthropomorphizing and animating, for whatever reason, cause us to lose sight of a lot of science. By using such words as "want" (the rock wants to fall), "need" (the tree needs to reach toward the Sun), and "feel" (the leaves feel the rush of the wind), we implicitly assume a degree of design that doesn't exist. Living things on Earth are driven by evolution to survive. As a result, they have developed many survival-related characteristics. Some are physical, such as protective shells, scales, or skin. Some are direct functions of bodies, such as the abilities to sense danger or food, to digest food, and to ward off illnesses. Some characteristics include innate and direct responses from the brain, such as the "desire" to nurture young, the impulse to fight for territory or mates, the ability to herd and to form community structures. For most animals, these responses are driven by hardwired functions in the brain and body, not by any form of reason. While we humans do all the same things, their origins are more complex because our levels of self-awareness and reason are greater. For example, when we are driven by two needs, such as hunger and the desire to read a good book, we can choose to read the book before we get something to eat.

Consider the universal use of the term "life cycle" to describe how a star changes with time (how it "ages"). The myriad aspects of this process are determined by the star's mass and, to a much lower degree, by its chemical composition. All stars begin their "lives" composed primarily of hydrogen and helium, with less than two tenths of one percent of their atoms composed of other elements. By comparison, oxygen, silicon, aluminum, iron, and calcium account for over 90 percent of the mass of the Earth's crust, while hydrogen is only the tenth most common element here and helium isn't even in the top fifteen.

Stars change chemistries as gravity causes fusion in their cores. Nowhere in their "evolution" is there any awareness of or response to their surroundings or indication of actions taken by choice. Furthermore, when stars explode, they send into space elements that would not otherwise exist, such as oxygen, silicon, aluminum, iron, calcium, and all the other elements necessary for complex life like ourselves to evolve. But the stars are not aware that their fusion processes will lead to life. All their activities are driven purely by physical laws.

One last issue often clouds our perception of natural events. It is our need to know "why" something happened. Suppose a friend of yours was killed by a drunk driver. I'd wager that you would spend a considerable amount of time during your mourning process asking why this "had to happen" and wondering, "If only. . . ." Such thoughts are part of how we try to make sense out of a sometimes senseless world. When natural events occur, we often do the same thing. An earthquake kills thousands, a volcano displaces hundreds, a tidal wave destroys a city, a monsoon washes away a country's entire crop along with countless homes and people. I think it is impossible *not* to ask "Why?" but if we give anthropomorphic or theistic explanations of natural events, we obscure our understanding of nature.

With all these sources of incorrect beliefs and the resulting misunderstandings of the natural world they create, how is it that we are still able to function so well in an increasingly complex world? Will we be able to continue to develop without a more accurate understanding of how things work? These are the questions I explore next.

4

Survival in a
Misperceived World

HOW WELL DID OUR ANCESTORS DO
WITHOUT UNDERSTANDING NATURE?

Stranger Than Fantasy

I find it especially remarkable how seemingly bizarre and counterintuitive the laws of nature are compared to even the wildest flights of human fancy. Reading or watching science fiction, one often gets the impression of great creativity on the part of the writers: "How do they think of those things?" For the most part, science fiction writers are able to extrapolate from existing technologies and scientific discoveries very effectively. By the time of Jules Verne, for example, submarines, rockets, telegraphs, and trains were already around as bases for his science fiction concepts. He used them to explore the implications of underwater travel, space travel, mass communication, and mass transportation.[1]

The more we learn about such fields as genetics, quantum mechanics, general relativity, astrophysics, and superstring theory,[2] the more fantastic things we learn about how nature works. Rarely, if ever, does a science fiction writer come up with an original, valid idea comparable in complexity and bizarreness to what scientists have discovered. This struck home for me a few years ago when I began to study superstring theory. While neither complete nor correct, it is beginning to provide us with a deeper insight into the nature of matter and space than any other theory we currently use.

The price we pay for pushing back the frontiers of scientific knowledge is the necessity of learning a level of mathematics the complexity of which far surpasses anything previously encountered. Once all that math is applied, however, it brings incredibly interesting results, including predictions of the presence of as-yet-undiscovered dimensions and a more coherent explanation of the properties of matter than our present piecemeal understanding. Superstring theories have the potential to lead us to a greater understanding of nature, but there is no evidence that science fiction writers are going to get there first.

You Can't Get There from Here

If you want to perform a socially acceptable but unnatural act, learn science. Science is socially acceptable primarily because its discoveries have provided more advantages than disadvantages to the human race. It is unnatural because it only developed during the past 800 years of the more than 100,000-year history of modern

---

[1] Verne's recently discovered novel *Paris in the Twentieth Century*, first published in English by Random House in 1996, is instructive in this regard.

[2] Superstring theories attempt to describe the fundamental nature of particles and how the forces in nature (gravitation, electromagnetism, and the weak and strong nuclear forces) are related.

homo sapiens.[3] Social acceptance of scientists is often qualified, however. My own experience is that we scientists are generally tolerated in social situations with nonscientists, but unless we're studying something really exciting, notorious, or exotic, most people aren't interested in our unnatural work. For example, there is a noticeable difference in how people respond to me when I say I'm a scientist ("Oh, that's nice") and when I say I'm an astronomer[4] ("You are, really? I always wanted to be an astronomer, but I couldn't do the math. I've had a telescope for years. Do you study black holes? I've always wondered . . .").

The unnaturalness of science becomes apparent to students studying any of its disciplines for the first time. Take, for example, the experience of physics. The first year of college physics courses is usually very traumatic. In the first place, you have to brush up on, or learn, the appropriate mathematics. This is already unnatural. Basic math like trigonometry and geometry has been around for less than 2,500 years, but calculus, upon which modern science rests, was first developed only in the latter half of the seventeenth century by Isaac Newton and Gottfried Leibniz, a German mathematician working in France. Even today, the vast majority of people on Earth know no more than rudimentary algebra and geometry, if that.

My own experience with college physics wasn't unusual. I began by simultaneously taking introductory calculus and calculus-based physics courses. Stumbling through these alien forests filled with unspeakable horrors, I bumped into many trees, tripped over many roots, got lost in the underbrush, and was frequently bitten by the common-sense bug that infests our expectations of the natural world. To make matters worse, I was well into my first year's courses before I realized how much the two realms of math and physics overlap. Con-

---

[3] I think that most historians of science would make the point that science began in ancient Greek culture three thousand years ago, and I would agree. However, science as a process (see chapter 1) didn't blossom until the time of Roger Bacon (1214–1294) and didn't come into full flower until the Renaissance began in the fifteenth century.

[4] Assuming they don't misunderstand and assume I'm an astrologer—see chapter 2.

TABLE 4.1 *Three Ways to Describe How I Drive*

Position	Velocity	Acceleration
Everyday My car is 10 miles north of Bangor, heading south on I95.	I am traveling at 65 miles per hour.	I just accelerated to pass another car.
Algebra My car is 10 miles north of Bangor, heading south on I95.	The change in my car's position divided by the time it takes to travel that distance in a certain direction is my car's velocity (65 miles per hour south). This assumes that the rate of change of my car's position is constant over the time interval in question. Otherwise, we have to talk about average velocity.	The change in my car's velocity divided by the time it takes to make that change is it's acceleration. Again, this definition assumes a constant rate of change of velocity. Otherwise, we have to talk about average acceleration.
Calculus My car is 10 miles north of Bangor, heading south on I95.	The change in my car's position at any instant (first derivative of position with respect to time) is its velocity (presently 65 miles per hour south).	The change in my car's velocity at any instant (second derivative of position with respect to time) is my car's acceleration.

sider, for example, the everyday concepts of an object's position, velocity, and acceleration. We learn what these words mean from our common use of language. They each pretty much stand alone in normal conversation (see table 4.1) and are interconnected only slightly more rigorously in algebra. However, only when calculus is employed is the larger picture revealed.

The use of everyday language is intuitive, if imprecise. Algebra introduces precision, but the terminology is cumbersome and often requires many caveats. Calculus results are precise, concise, and elegant. In general, the more powerful the mathematics, the more comprehensively it describes the physical world.

Students who move farther and farther away from the familiar world into the Byzantine realm of science discover that many common-sense ideas are inconsistent with the equations they learn to solve and to believe. In that sense, science is unnatural. For example: throw a ball straight up in the air and catch it when it comes down. At the top of its travels, the ball stops. What is its acceleration at that point? As mentioned earlier, most people reason that since the ball is not moving at the top of its motion, it can't be accelerating there, making the answer zero.

The correct answer is approximately 9.8 m/sec^2 or 32 ft/sec^2, the acceleration due to the gravitational attraction of the Earth. This is the same acceleration the ball has throughout its entire flight. To understand why acceleration at the top is not zero, note from the calculus row in the previous table that acceleration is the instantaneous change of an object's velocity. Right before the ball reaches the top of its path, its velocity is upward. At the top, its velocity is zero, so in that moment it has changed velocity, or accelerated. Between the instant it is at rest at the top and when it starts falling, it again changes velocity, or accelerates. The position at the top represents the time at which the velocity changes from upward to downward. But changing velocity can only occur if the ball is accelerating throughout the transition. For the ball to have zero acceleration (as well as velocity) at the top, it would have to stop there permanently.

Another way to look at the same problem is from the perspective of Newton's law of motion, which states that the force acting on any object is equal to its mass times its acceleration. The force on the ball is the gravitational attraction from the Earth and the ball's mass is just the number of particles it contains. When you throw the ball up, both the force from the Earth and the mass of the ball remain the same. Therefore, throughout its flight, the acceleration on it must remain constant.

No one is exempt from the conflict between intuition and the laws of nature. I began this chapter by noting that science is unnatural from our everyday perspective, but science is actually a much more accurate representation of nature than our common sense. From that point of view, everyday or common-sense thinking about nature is what's really unnatural.

Math teachers supply us with many of the tools we need to correctly comprehend science. Without mathematics, "you can't get there from here." But even with these tools, we must let go of many beliefs, both common sense and learned, before we can really appreciate how nature works. It's worth noting that mathematicians are continually discovering new mathmatics. Indeed, some of the new math being developed even as you read this provides the key to uniting the laws of nature on the largest scale, cosmology, with the laws on the smallest known scale, that of elementary particles.

Even when you have the essential mathematics in hand, the terror of learning physics doesn't abate. The next hurdle I found was learning which equations to use under which circumstances. This ability only comes with long, painful practice because there are so many equations and so many circumstances. That's as far as most students taking physics courses ever get in dealing with true natural laws. Plug and chug. Students at this stage are encouraged to check their answers to see if they are realistic—are the units appropriate? Are the numbers physically reasonable? Most students don't even do this amount of work. Your doctor probably didn't check his or her answers in this way when he or she took the two required physics courses to get into medical school.

From the beginning, part of learning to use equations is learning the scientific definitions and realities of terms you think you already know, like "position," "velocity," "acceleration," "time," "energy," and "force." Consider time. Chances are that if you were asked to compare the rates at which time flows (at which clocks tick) for someone standing on the ground and another person in a moving car, you would expect that their two clocks tick at equal rates. They don't. Time travels more slowly for something that's moving than for something at rest. When the moving person stops and the two people compare themselves, they will find that the moving person aged more slowly than the person at rest.

After learning the correct meanings and behaviors of words you think you know, you must learn the physical meanings of some new terms, like "potential," "vector," "field," "entropy," and "heat capacity." Even understanding what the terms mean and knowing how to choose the right equation isn't truly enough to understand nature. The next stage is learning to manipulate the equations to determine things that the equations don't tell you directly but *imply*. For example, an equation may be set up to tell you how far something moves, given that you know its speed, its acceleration, and the length of time it is moving. But instead suppose you are told how far it has gone, its speed, and its acceleration, and are asked to find out how long it traveled. To do this, you would need to manipulate the equation you started with—essentially create another one you may have never seen before. The solution for the travel time is implicit in the equation for distance traveled.

If you haven't taken a math or physics class in which such manipulation is taught, consider this analogue: you take classes that involve descriptive concepts and lists of facts. You memorize information for an exam and then the teacher asks a question that requires you to reverse a list or use two facts together to draw a logical or plausible conclusion. For example, in my astronomy class I teach the order of distance of the planets from the Sun. I also teach Kepler's third law, which states that the closer a planet is to the Sun, the faster it orbits. Put another way, the closer a planet is to the Sun, the shorter its year. When asked which of the following planets is closest to the Sun— Jupiter, Saturn, Venus, Earth, or Neptune—most students will give the correct answer (Venus). When asked the question: "Which of the following planets, Jupiter, Saturn, Venus, Earth, or Neptune, has the longest year?" many more give the wrong answer. The correct answer is Neptune. Getting it requires reversing the statement of Kepler's third law (from "closer planets have shorter years" to "farther planets have longer years") and then using this result with the given list to find the most distant planet.

Once you think you have the basic techniques and concepts of physics down, you are brought up short by the plethora of specialized topics: classical mechanics, quantum mechanics, electromagnetism, statistical physics, thermodynamics, elementary particles, general relativity, solid state physics, nuclear physics, atomic physics, plasma physics, surface science, astrophysics, geophysics, biophysics, optics, chaos, and fluids, among others. All other mature fields of science have many specializations too. Furthermore, each of these specialized topics is divided into numerous subtopics, all with increasingly complex mathematical structures. It is necessary for physicists to specialize in one such subtopic or another.

As a field of science matures, it develops more and more techniques for solving problems. Therefore, even many advanced physics students stumble when they face the challenge of trying to choose the correct mathematical procedure to use. Indeed, I have seen graduate students brought to their knees by their inability to find the starting point for solving advanced problems.

While physics is known to be "a bear," understanding astronomy at the same depth is actually harder because you have to combine a diverse range of math and physics in order to comprehend what is happening in the cosmos. For example, to understand the Sun, you have to understand substantial amounts of thermodynamics, statistical mechanics, electromagnetism, classical and quantum mechanics, nuclear and atomic physics, and plasma physics, at the very least. Every semester, without fail, students take my introductory astronomy course, find the subject enchanting, decide they want to study it further, and then experience a terrible shock when they find out that learning anything more about it formally would require years of math and physics first.

How Our Distant Ancestors Survived

I hope the foregoing discussion gives you insight into why so few people understand or study science and why it has taken so long in the history of *homo sapiens* for science and the scientific method to develop. This progress has occurred primarily over the past six hundred years. Prior to that, from the time of the first sentient creatures on Earth, understanding of the cosmos was based almost entirely upon experience and common sense. As discussed in the preceding chapters, our 20/20 hindsight has revealed that most of what our distant ancestors thought about the natural world was wrong. This raises the question of how they survived over innumerable generations in their state of cosmic ignorance. It is my contention that they made it by the skin of their genetic teeth and that had they remained in a nonscientific state of mind, our lives today would be much less comfortable and sophisticated.

First, consider the physical well-being of our ancestors. Just how good was their health prior to the development of modern medicine and hygiene? It's probably just as well that we can't see the universe of bacteria, other microscopic organisms, and viruses with our unaided eyes. I think we would be quite intimidated by both the range and the number of tiny life forms[5] that inhabit just about every corner of the Earth, including ourselves.

We have constructive relationships with some of these organisms. For example, we have symbiotic relations with a variety of bacteria, such as those that normally infest our intestines. Members of the lactobacillus family of bacteria help us digest glucose, for example. The yogurt I just ate was a good source of lactobacillus acidophilus, useful in digesting dairy products. Lactobacillus also creates lactic acid in the human vagina, which inhibits the growth there of many types of harmful bacteria.

---

[5] For the sake of simplicity, I will include viruses in such expressions as "life forms" even though they are not technically alive.

Conversely, many microorganisms make us ill. Our bodies have evolved an impressive array of natural defenses against many of them. Such built-in defenses have been essential to the survival of virtually all life forms. Nevertheless, as you know from everyday experience, we must actively deal with many illnesses in order to remain healthy and to prevent permanent damage to our bodies. Without such help, our ancestors suffered considerably more than we do, especially the children.

All children go through a variety of common illnesses. My boys got their share of ear infections, strep throat, conjunctivitis, diarrhea, and elevated temperatures as their bodies struggled to fight various microorganisms. In each case, we would find out the appropriate treatment and help their natural defenses.

Now consider what happened to children prior to the nineteenth century, who lacked the resources of modern medicine. Ear infections frequently resulted in partial or total deafness. This hearing loss in turn frequently delayed speech and language development. In many cases, ear infections led to meningitis, an inflammation of the membranes that cover the brain and spinal cord. It is potentially life threatening, and survivors often have permanent hearing or vision problems. The vast majority of children develop ear infections. Therefore, the vast majority of our ancestors who survived them had some degree of hearing impairment, at least. Many also had delayed language development compared to modern, healthy children.

Strep throat is a common illness caused by a streptococcus bacteria. It is easily treated today with antibiotics. Our ancestors who got strep throat had some serious aftereffects with which to contend. The most notable was rheumatic fever, which can cause arthritis and heart problems. Strep also causes kidney inflammation and scarlet fever. Therefore, many of our surviving ancestors had arthritis and weak hearts from this illness.

Conjunctivitis is a group of diseases with a variety of environmental, bacterial, and viral causes. As you probably know, it causes the eye to turn pink and is accompanied by a variety of temporary symptoms. Our bodies' defenses can overcome some types of conjunctivitis; others require antibiotics. Untreated, some conjunctivitis can lead to permanent vision impairment or even blindness.

Diarrhea is caused by a variety of viruses, bacteria, parasites, food allergies, and medicines. Prolonged diarrhea often causes headaches, fever, nausea, and vomiting in a process of dehydration that can lead to death. Indeed, diarrhea is one of the most common causes of child mortality even today. It is common where sanitation is poor and where food and water supplies are contaminated. Fortunately, in developed countries these conditions are rare, but they are still common in many developing nations and were the rule throughout the world until the nineteenth century. Indeed, until adequate sanitation in the twentieth century, virtually all classes of people everywhere were susceptible to diarrhea-related illnesses. Its worth noting that these illnesses grew more widespread as more people began living together, a point to which I shall return shortly.

High fevers result when our bodies automatically raise our temperatures to help kill temperature-sensitive invaders. The problem is that high temperatures can also cause permanent damage to our bodies and even death. Keeping a fever down while the body fights illness by other means is therefore often a desirable approach. It was not an option in most places until the mid-eighteenth century, when willow bark[6] was first widely used to lower fevers. Willow bark has the significant side effect of causing extremely painful stomach distress. It wasn't until 1899 that the active ingredient, salicin, was first converted into a relatively well-tolerated compound, acetylsalicylic acid (aspirin). Prior to the twentieth century, therefore, the consequences of high fevers were widespread. Most people suffering from them either died or had permanent physical impairment.

Aspirin brings us to the issue of pain. Among those people in earlier centuries unfortunate enough to survive childhood illnesses, pain, sometimes excruciating, was a part of everyday life. If you don't believe that, imagine what your life would be like if you had a

---

[6] Some tribes around the world with access to willow trees had known about the bark's fever-reducing (and pain-relieving) properties for hundreds, if not thousands, of years. Shamans or medicine men also knew about and dispensed a wide variety of other medicinal aids. Some helped, some didn't, but no one back then understood why or how.

toothache[7] that persisted for weeks and months. Unless the tooth was extracted, which was also extremely painful without anaesthetics, you were stuck with the pain. (Early pain relievers included alcohol, opiates, and mandrake root. Ether, the first modern anaesthetic, was first used in 1842.) Pre-anaesthetic surgeons used to pride themselves on cutting off limbs quickly enough so that the patient didn't die from the pain and loss of blood before the surgery was done.

Even if you did survive an operation prior to the use of anaesthetics, you would probably die of sepsis (the spread of bacteria from an infected area). It was only in the 1860s, when Louis Pasteur proposed that living microorganisms caused illness, that Joseph Lister showed that sterilizing everything in the medical setting prevented sepsis and saved lives. (Yes, Listerine was named in honor of him.) Without anaesthetics and antiseptics, living was often more painful than dying.

Freedom Through Vaccination

Another whole class of illnesses deserves attention for being brought under control by scientific advances, namely the diseases for which vaccinations are in widespread use. In the United States, for example, it is common to immunize children against many diseases that ravaged the ranks of our ancestors. Vaccines prevent hepatitis B, diphtheria, measles, mumps, German measles, pertussis (whooping cough), polio, tetanus, rotavirus, haemophilus influenza type b (HIB b), chicken pox, and a common pneumonia bacteria. All these diseases terrorized our ancestors, who, admittedly, didn't know their causes but certainly knew their effects. While there are innumerable stories about all these illnesses throughout history, let's consider just two examples.

---

[7] Toothaches are perhaps a bad example, since they are quickly becoming a thing of the past. I had dozens of cavities as a child, along with the resulting toothaches. My children have had few, if any. If the pain associated with a toothache is alien to you, imagine prolonged suffering with the worst throbbing pain you have felt.

THE TERROR OF TETANUS

Tetanus or lockjaw is a disease for which most of us are vaccinated in childhood, with booster shots given every decade. As a result, our major concern about most minor cuts, splinters, insect bites, and burns is usually cosmetic. However, all such wounds are sources of *clostridium tetani*, which causes tetanus. This bacterium is a common component of everyday soil and is also found in animal intestines (including our own). It does not get from there into our bodies, however; it enters through our skin. The results of untreated infection are often fatal, since the tetanus toxins cause muscle spasm or contraction (hence the name "lockjaw"). The entire muscular system is affected, and breathing eventually becomes difficult and often impossible.

Prior to the early 1890s, there was no preventative medicine against tetanus. Therefore, the disease was a hazard to anyone who got cuts, splinters, bug bites, or burns. This, of course, was everyone. Have you noticed that paintings and photographs of beach scenes from the nineteenth century and earlier show people fully clothed? That style was a result of Victorian sensibilities, but it had the added benefit of protecting wearers from insect bites, which back then were often lethal. Tetanus was also a major cause of death for soldiers injured in combat.

POLIO—SUMMER NIGHTMARE

Today, summertime in the developed world is often a season of relaxation, rejuvenation, play, travel, and lots of outdoors activities. Prior to the development of the Salk polio vaccine in 1955, though, summer was polio season. It was a time of terror, when parents knew they stood a good chance of losing children to this disease. Polio kills nerve cells, which causes muscles to atrophy. When nerves controlling breathing are affected, victims stop breathing and die. Those who survive the initial onslaught of the disease are permanently crippled by it. Franklin D. Roosevelt, to name possibly the most famous example, was struck with the disease in 1921, leaving him unable to walk unassisted. Many survivors had to stay in iron lungs in order to breathe. I was born shortly before a vaccine became available, and I still vividly remember pictures of people in those enormous contraptions.

Frequent polio epidemics occurred throughout the world, causing people to leave infected areas in droves, until the Salk and Sabin vaccines became widely available. These vaccinations changed how people viewed social activities, especially during the summer. So when you go to the beach wearing a sexy bathing suit, remember that there is a very good chance that science is helping to keep you alive and well.

Death by Society

As our ancestors began living together in increasing numbers in pre-scientific times, their close proximity created problems that often led to illness and death. Science and technology have overcome many of these. Consider life in early cities. Two thousand years ago, the relatively sophisticated city of Rome used lead pipe to plumb wealthy homes and large waterways. The lead was procured by smelting, the melting and fusion process by which metals are extracted from ore, which produced especially toxic emissions—an early form of human-made air pollution. Lead was also used in a variety of pottery glazes, in linings for wine containers, and even to whiten the skin in the form of lead acetate. All of these uses caused lead poisoning, symptoms of which can include tiredness, appetite loss, pale skin, changes in behavior, vomiting, stomach pain, headaches, confusion, seizures, weight loss, numbness or tingling in arms or legs, memory loss, and diminished mental development in children.

To verify the problems lead created in ancient Rome, modern researchers have made wine in the way described by period Roman writers, using lead-lined vats and containers. When tested, the resulting wine had very dangerous levels of lead. Likewise, lead in makeup worn by Roman women is known to have leached through the skin and caused poisoning. While lead may not have caused the downfall of Rome, it certainly didn't help. Lead-based paint was used in households in the United States until the middle of the twentieth century, putting children who licked or ate chips of it at risk. Lead was an early additive in gasoline, so it got into the atmosphere and endangered everyone until it was banned in the 1970s. Since science has quantified the dangers of lead in recent decades, its uses have been greatly restricted.

Another example of how population density can lead to severe health problems is associated with the emissions from early heating systems. Prior to the development of heating techniques such as solar, electric, natural gas, and high-efficiency oil furnaces, heating was done primarily by burning wood and coal. Such fuels burn very inefficiently, emit a variety of chemical compounds that cause breathing problems, and deposit grime wherever the smoke settles. As a result, respiratory problems were endemic in cities until the second half of the twentieth century.

Cardiff, Wales was a steel-producing town when I lived there as a graduate student in the 1970s. Vast quantities of coal were used in the smelting process in factories on the outskirts and to heat many of the flats (apartment dwellings) throughout the city. As a result, Cardiff was perpetually grimy and smelly. One of my good friends, also studying general relativity there, became very ill. For several weeks he could barely function. His mother, a physician, finally diagnosed the problem as carbon monoxide poisoning. It turned out that the coal heater in his flat had developed a leak, allowing the odorless, colorless gas to seep into his bedroom.

Throughout the world, there are still enormous air pollution problems, created in large part by motor vehicle exhaust. The pollution was much worse when gasoline contained lead and exhaust vapors were vented directly into the atmosphere through exhaust pipes. When the hazards of unprocessed emissions were made clear in the 1960s, laws were passed in the United States that banned lead in gasoline and required that exhaust be cleaned before it was emitted. Because the chemistry of the exhaust gases and a large variety of chemical reactions were well understood, scientists and engineers were able to develop catalytic converters that change the gases into less harmful compounds.

Suppose cars had never been developed. If we still used horses to get around, the solid waste and air pollution problems would be staggering. In the nineteenth century, cities like New York were already ankle-deep in horse manure, and the air contained large amounts of horse-generated methane and other noxious gases. We would have much greater problems than we do with car exhaust if we still used horses for transportation and had to deal with hundreds of millions of them and their "exhaust."

EPIDEMICS

Large groups of people who lived together in cities or other close groups such as armies in prescientific times also died as a result of fast-spreading and devastating diseases. Epidemics were often due to poor sanitation, overcrowding, rodents, and airborne carriers like fleas and mosquitos. Major epidemic diseases included bubonic plague, smallpox, typhus, influenza, whooping cough, diphtheria, dysentery, measles, malaria, yellow fever, scarlet fever, cholera, dengue fever, and polio. The Great Plague of 1665 in London killed over 20 percent of that city's population. Smallpox introduced by European colonizers wiped out over 95 percent of the Native Americans living along the Connecticut River. Likewise, it killed some 200,000 Peruvian natives in the 1520s. Many of these epidemics spread throughout entire countries and continents; when they become pervasive on such large scales they are called pandemics. The 1918 influenza pandemic killed over half a million people worldwide. We are in the middle of an AIDS pandemic as I write this book.

Interestingly, in 1665–66, an epidemic of the plague closed England's Cambridge University and forced Isaac Newton to retreat to his home at Woolsthorpe, in Lincolnshire, England. Sitting out the plague on his own, he developed the mathematical theory of gravitation (Newton's law of gravitation), showed that white light is actually a combination of all the colors of the rainbow (initiating our present understanding of color and optics), and did some of his pioneering work on calculus, thereby helping to jump-start the scientific revolution that eventually helped control many diseases.

Not surprisingly, children have been more susceptible to illnesses we now control than adults. Therefore, infant mortality is used as a measure of how successful medicine has been at preventing and controlling illness. In the 1780s, one writer reported that half the children in Paris died under age 2. At the same time, Thomas Percival, writing in England, noted that half the children born in Manchester, England died before they were 5. By the 1860s, infant mortality ranged from about 14 percent in Sweden to 22 percent in France. Today these countries, along with the rest of Western Europe, have infant mortality rates of less than 1 percent. Science has been crucial in sustaining and improving the quality of life for children and everyone else on the planet.

Other Contributions of Science to the Quality of Life

How well did our prescientific ancestors fare in areas besides medicine and health? Again, we can get a perspective by considering how science has helped make life much more comfortable and productive. I will discuss engineering products, rather than science, in this section. It's important to realize that early inventions like the steam engine, camera, and telegraph were built before anyone truly understood the scientific principles by which they worked. All that has changed. Today engineers can create the vast majority of new technologies only because they understand the underlying science of the products beforehand. Indeed, we now know how many things work from the atomic level on up.

SPEED

Let's consider just one example, namely our ever-improving ability to travel. It took the *Mayflower* 64 days to cross the Atlantic in 1620. In 1854, the fastest crossing of the Atlantic was made in 13 days, 20 hours. Science and the technology that resulted from it took off in the twentieth century; understanding of thermodynamics, friction, combustion, and metallurgy led to development of better engines and structural designs, which compressed travel time dramatically. The S.S. *United States* (the fastest transatlantic ocean liner) took just over 3 days and 10 hours to cross the Atlantic in 1952.

The same evolution took place, although more rapidly, in air travel. It took Charles Lindbergh 33 hours, 29 minutes to cross the Atlantic in 1927. In 1950, the Lockheed Constellation passenger plane took 12 hours to fly from New York to London. Today, normal passenger flights between New York and London take 6 hours. The supersonic Concorde made the trip in just under 3 hours. Furthermore, our pilots know where their aircraft is located to within a few yards and what their arrival time will be to within a few minutes, traffic and weather permitting. This accuracy is due entirely to the application of science to the field of navigation.

Such high-speed transportation and instantaneous communication have brought people closer together. Virtually anyone in the developed world can contact anyone else within a matter of minutes. Our technology has broken down many of the barriers that once served to help us define communities. We can debate whether or not this is a good thing. In any event, we are now in contact with more people and can witness and participate in a wider range of events than ever before. Unfortunately, rapid transportation also means rapid spread of diseases and other ills, such as terrorism and insect pests, around the world.

THE . . . OLD DAYS

Periodically we all experience waves of nostalgia for a simpler, friendlier time as envisioned by painters Norman Rockwell and Thomas Kinkade. Those idealized eras never existed, except possibly for the upper, moneyed classes. True idealizations are misconceptions we create by comparing what we perceive as a hectic, fragmented existence with a slower, more local way of life focused on survival. I have tried to make the case that our ancestors did not live idyllic lives. On the other hand, I personally don't think that our present social, scientific, and technological conditions are ideal. Rather, I believe that we are in a period of unprecedented transition. Transitions are extremely disconcerting, and in today's world we rarely have time to stop and take our bearings before some new technological, medical, social, economic, or political change drives society in another direction.

I believe that, barring a global catastrophe, we will continue on the high-tech road we presently travel. This means that we will continue to understand better how nature works and to find applications for this knowledge. Further, I believe that we need to understand the effects of the changes that are occurring to be better able to decide as a society where to go and what to do. For example, the Internet is presently a rapidly changing, nearly universal presence in our lives. Some aspects of the Web are very good: it provides much accurate information and many constructive, differing viewpoints, more than were ever available to individuals using conventional media. You can often do more research in a day by "surfing the Web" than you used to be able to do in a week at the library. However, some aspects of the Web are very bad: it also provides more disinformation and destructive viewpoints than ever before.

We can help guide the evolution of the Web by deciding what we do and do not want from it and working to ensure that we maximize its benefits and minimize its disadvantages. This might include imposing stiff penalties on people who create computer viruses or distribute child pornography. It might mean using software filters so that certain sites are not available on individual computers or through major servers to protect children. The point is, we are in a technological and scientific transition and we need to understand it to make it work best for everyone.

Although science is still alien to most people and contrary to our common sense, it enables a vast array of improvements in our lives as well as a universal (as compared to an individual) way to comprehend the universe. Science is an impersonal, objective, public discipline compared to art and other creative activities, which express personal, subjective, private creativity. However, science is accomplished by personal effort, often for subjective reasons, and often in private. The difference between science and art is that science forces consensus—scientific theories are either accurate or inaccurate, whereas artistic creations do not require consensus and, indeed, rarely create it. The difficulty in adopting a scientific mentality and working within its often restrictive guidelines is one of the reasons most humans have not made science a big part of their existence until recently.

Why Do We Have Common Sense Rather Than Scientific Sense?

Why did our brains evolve to perceive the natural world so differently from the modern scientific perspective? I remember thinking about this as a teenager. At that time I believed that a higher intelligence micromanaged our lives. I reasoned that we had such screwed-up natural perceptions because that higher intelligence wanted us to attain certain levels of social, political, and ethical maturity as a race before we could be allowed to understand and use the forces of nature. Keep in mind that this was at the height of the Cold War, when the fear of nuclear holocaust was on everyone's mind. Perhaps we had gotten the scientific knowledge too soon, I reasoned, and we were still so immature as a race that we would blow ourselves up.

I must admit that such thoughts about mass destruction made me question the higher intelligence idea. If that intelligence was really "higher," it clearly wouldn't allow us to use the bomb again. But if we believed that and therefore didn't keep a supply of bombs, could we really trust the other side not to use them? Most people thought not, so the stockpiles of nuclear weapons grew. Trying to retain my own belief in a higher power, I reasoned that perhaps the Cold War was a test to see if we were truly ready to have real insights into nature. You can see how convoluted such thinking becomes. Remove the higher intelligence from the mix and the same reality becomes much easier to explain: we had unlocked some of the secrets of nature, and their potential for destruction was forcing us, the human race, to grow up.

So why, then, did humans in general evolve different reasoning processes and different "common sense" than those used by scientists studying nature?[8] The answer is that we evolved to survive. I do not believe that there was ever any "intention" by any external power that our brains should necessarily evolve self-awareness, logical thinking, the ability to communicate, or the ability to override our instincts.

---

[8] I explicitly say "studying nature" because when scientists aren't thinking about their field, they often use the same incorrect reasoning and common sense as everyone else.

These and related human characteristics came with the complexity that developed from more basic pieces of our neural system.

Different parts of our brains evolved to help us perform different tasks. We have very limited control of autonomic functions, the things that our brains do for us automatically. For example, our brains control how fast our hearts beat. We cannot stop our hearts or consciously regulate the actions of most of the other organs in our body. The closest we come to controlling our continuous internal body functions is with breathing: we can deliberately override the normal rate at which we breathe. But we can't stop breathing altogether. If we try, we pass out, and our brains take over again, forcing us to inhale and exhale.

Above the autonomic part of the brain is a set of complex systems that enable us to store information and respond to stimuli. Our memories, though still poorly understood by science, have been shown to be built of a hodgepodge of different systems. We don't remember everything about anything all in one place in our brains, in the way that photographs record an entire scene visible through the camera lens. Rather, our brains piece together memories from a variety of storage areas. For most of us, this recall process is imperfect but very rapid. Fast mental response was essential for survival but perfect recall wasn't necessary.

One dimension of responding to the environment was to deal with danger. Rapid recognition of a threat had to be followed by rapid decision making: fight or flee? Often our ancestors had only a matter of seconds or less to come to the "right" conclusion. Rapid decision making is often inconsistent with the consideration of many details or issues related to the problem at hand. Even after the pace of life slowed down and we had the luxury of time to think, we still had minds that were tuned to take a few pieces of information and quickly come to some decision based on them. We have to *think* to overcome this hardwired aspect of how our minds operate.

An even earlier part of our response to the environment came from the need for our ancestor species not to fall out of trees. Arboreal animals that slip and fall only have a very few seconds to stop themselves. Three seconds after they start falling, they are going downward at over sixty-five miles an hour. Our ancestors had to either grab something in less than three seconds or be hurt or killed, depending on how far they fell. Therefore, our brains evolved to respond rapidly to falling, rather than to analyze how our speed changes as we fall—there was no time for that! Fortunately for us, by the time we were capable of surviving very challenging environments, the structures in our brains had evolved to the point where other thought processes were also possible.

Crucial in our evolution was the development and refinement of the senses, which evolved to help us identify crucial things like danger, food, and potential mates at different distances. Touch, taste, and acceleration give us feedback about our immediate environment. Smell, sound, and heat are useful for objects nearby, while sight provides information about things from here to infinity. Again, our senses evolved to help our survival: they did not have to be perfect; they just had to work. Our ancestors often did not have time to process the information their senses provided very deeply before they responded. However, the evolution of our brains gave us that ability, and now we live at a time when we have opportunities to use it.

Nowhere in the evolution of human awareness or reasoning is there a need for experimentation. Curiosity, yes; testing beliefs, no. If something worked the first time, our ancestors stuck with it, and if something failed or hurt the first time, our ancestors avoided it. But experimentation is the heart and soul of science. The need to test ideas had to be discovered by trial and error together with deep processing. The first approach to understanding the cosmos used *just* reasoning, like that of the ancient Greeks. With few exceptions, such as Archimedes, the "thinkers" believed they could deduce how nature worked using pure thought.

The trial-and-error phase of developing science came primarily in the Renaissance, when people like Leonardo da Vinci had the resources and the time, over and above that needed for day-to-day survival, to make things. Some inventions worked, some didn't. After each failure, da Vinci examined the causes. Such reasoning led to new ideas, and those led to further trials. This seems obvious today. However, as late as the Renaissance, craftsmen who could actually build things, as opposed to "theorists," were held in relatively low esteem. The process of slowing down and looking more deeply at what we sense and what we think we know is quite a recent development. As Lewis Wolpert said in *The Unnatural Nature of Science*, science is a self-aware process, while common sense is not. You might even say that studying science is a leisure-class activity.

KNOWLEDGE IS POWER

The crucial elements of our movement into the scientific age are that, as a species, we have learned how to use the scientific method; we have acquired a body of knowledge that reflects at least shadows of reality about nature; and we have applied what we've learned to create things. All the advances that science has brought have been carried out and understood by relatively few people. Any scientific field has at most thousands of experts, and the world contains billions of people. The vast majority of the rest of us profit from the results of scientists' work without ever fully understanding it.

I am not advocating that we all go out and learn everything about science or even about one field, such as astronomy. There is a big difference between such an unrealistic expectation and what I think is important, namely that we try to unlearn incorrect beliefs and avoid adopting more in the future. There are several reasons why this is worth doing. First, the more scientifically correct information we believe, the better we are able to judge the impact of a wide variety of issues on our lives. That impact is often financial and often affects our "way of life." Consider, for example, the debate about global warming. This phenomenon is caused by gases in the air called greenhouse gases that absorb heat from the Earth that would otherwise radiate out into space. If the heat remains in the air, it gets warmer and can alter global sea levels, weather patterns, and farming conditions, among other things. Such changes can, in principle, have dramatic effects on the Earth.

The issues related to whether global warming is really occurring are complicated by the fact that the Earth goes through a variety of natural cycles of warming and cooling. Although we see a warming trend right now, can we be sure that it is due to additional greenhouse gases put into the air as a result of our activities? The immediate importance of making a correct decision about global warming comes down to money. If our technology and lifestyles are at fault and we have to change them, then we are going to have to pay for it. For example, while automobiles in the United States now must comply with emission standards for pollutant gases, vans, trucks, and vehicles with two-cycle engines have been exempt from the restrictions and are dumping thousands of tons of the same gases into the air every year. There will be costs associated with cutting back on emissions from these vehicles. How much will we reduce greenhouse gases—and is it worth spending the money?

While many of us have "gut" feelings about such issues, we can do better. We can learn to think more scientifically and more critically, and thereby come to more accurate conclusions. If you live in Detroit and your livelihood depends on making trucks, your work environment may affect your belief in the reality of greenhouse warming and the significance of vehicle emissions to it. If your boss tells you that more studies are necessary to determine what is true, you might well find that enough reason to defer making any changes. But if you learn to think critically—Does my boss have a bias that could affect what he is telling me?—you can investigate other sources and get enough information to come to a more knowledgeable, valid conclusion and participate more constructively in the process of change that society is undergoing.

Let's now take a look at some of the reasons why this is easier said than done.

5

Breaking Up Is Hard to Do

MISCONCEPTIONS ARE
HARD TO REPLACE

Full Moon on the Rise

Probably the most emotionally laden object in astronomy is the Moon, and the most emotional issue related to the Moon is whether it affects our behavior when it is full. The vast majority of people believe that strange things are especially prevalent at the full moon. There have been literally hundreds of studies about the Moon's effects on assaults, kidnappings, domestic violence, shooting incidents, stabbings, homicides, accidents, depression, anxiety, suicides, prison violence, psychiatric patient admissions, 911 calls, emergency-room visits, natural disasters, human-made disasters (like train wrecks), alcoholism, casino activity, illegal drug use, and drug overdoses. A few of the studies show a measurable increase in these occurrences at the full moon over any other lunar phase. Interestingly, for each study that shows a correlation between some particular type of activity or event and the full moon, several more-thorough studies show no correlation whatsoever.

A scientific decision whether to believe that the Moon affects our lives and activities has to be based on statistics. There is no cut-and-dried answer. The trouble is that we humans are very poor at analyzing and interpreting statistics without a lot of training.[1] This is another instance where common sense and valid conclusions are at odds. The human shortcoming related to statistical analysis is made worse with topics like the effect of the full moon because our thinking about them has a cultural or emotional element. So, for what it's worth, statisticians who have reviewed all the studies related to the effect of the full moon have found no statistically significant correlation between it and various activities and events.

"But that's only statistics. Statistics can be manipulated to give you any results you want."

---

[1] Consider, for example, Simpson's Paradox. Suppose you need surgery and can have it done at St. Mary's or St. Jude's hospital. You ask your doctor, who tells you that the mortality rate (death within six weeks as a result of surgery) at St. Mary's is 37/1000, while at St. Jude's it is 30/1000. At which hospital should you have your surgery? Your "intuition" tells you St. Jude's, of course, but you may well be wrong.

Digging further, you discover that people like you who are in good condition prior to an operation die from surgery at a rate of 12/1000 at St. Mary's and at 15/1000 at St. Jude's. People in poor condition die from surgery at a rate of 42/1000 at St. Mary's and at 46/1000 at St. Jude's. Now wait a minute. How can St. Jude's, which has a lower overall death rate from surgery, have higher death rates for patients in good health and in poor health? By this result, you should have your surgery at St. Mary's even though it contradicts your intuition.

The resolution of this apparent paradox is the number of patients in good and poor health seen at the two hospitals. Of all the patients who have surgery at St. Mary's, more are in poor health than in good health. This raises the *overall* mortality of surgery patients at St. Mary's. Of all the patients who have surgery at St. Jude's, more are in good health than in poor health, which lowers the *overall* mortality of surgery patients at St. Jude's. You really should have your surgery at St. Mary's.

There is some truth in that statement. If you need absolute certainty in order to believe something, then you will always have reason to believe that there is a connection. If you have an emotional attachment to the belief that the full moon causes certain activities, there will probably never be evidence strong enough to convince you otherwise.

From both statistical and scientific points of view, the evidence about the Moon points to no correlation. Given that, the scientist in me is curious to find out why so many people believe the contrary. I suggest two plausible causes. First, many stories are passed down by word of mouth about the Moon. It is the Hydra of urban mythology. These tales are told to us from the time we are children. We hear many stories about the Moon's effects, often and from many sources. Under such circumstances, it's hard not to believe that they are true. Second, the Moon gets plenty of play in the media and in fiction. From werewolf legends to love stories, the full moon is alleged to have an effect on people. Even television weather forecasters often comment on the full moon.

Beliefs about the Moon's effects don't end in childhood. Many adults of all backgrounds and educational levels embrace them. For a decade, I worked as a volunteer emergency medical technician. As part of our training, we would spend three or four shifts a year working in a local emergency room. I tried to have my shifts in the ER coincide with the full moon whenever possible. On clear nights of the full moon, someone would usually remark, "Full moon tonight, we're going to be busy." And it often seemed that we *were* especially busy that night. Each case reinforced the belief that the full moon was causing a lot of activity, though the same number of cases on other nights just seemed like a normal workload. This effect is sometimes given the name "confirmation bias." Each event confirmed the prior belief of the effect of the full moon.

On cloudy nights of the full moon or when no one noticed the lunar phase, expectations about activity were not raised. The ER was typically no more or less busy than when the staff knew the Moon was full, but people didn't tend to notice that fact. Such personal experience as I had can be deceiving; to know whether lunar phases really affect ER activity, it is essential to get a much larger sample than a few nights a year. Hospital records show that there is no statistically significant increase in ER traffic on nights of the full moon compared to other nights (after accounting for seasonal changes and weather-related events).

Big Moon on the Rise

Another very interesting belief about the Moon that generates an emotional response is the conviction we all have that the Moon is bigger near the horizon than it is high in the sky. It may look bigger, but it isn't.

"Yes, it is!"

I respect this belief, but I'd like you to do the following experiment. Find out when the Moon is going to rise (you can get this information from a computer program like *Starry Night*, from the Web, from the paper, or from a weather forecaster, among other sources). Locate the Moon as it is coming up. Take an index-card-sized piece of cardboard outside with you and hold it at arm's length (measure the distance from the cardboard to your eye with a yardstick if you want to be really precise). Hide half of the Moon with the cardboard and then mark on the cardboard, as accurately as you can, where the two edges of the visible half of the Moon intersect the edge of the cardboard (thus measuring the Moon's diameter). Find something to do for at least three hours and then repeat the experiment, holding the cardboard at the same distance from your eye, when the Moon is high in the sky.

Now that you have discovered on your own that while the Moon *appears* bigger near the horizon, it actually covers the same amount of area of the sky the entire time it is up, let's consider why we all believe that it changes in size. Many explanations have been proposed,[2] but I

[2] For a more detailed discussion of this issue, please see "New Thoughts on Understanding the Moon Illusion" on the Web at www.griffithobs.org/ IPSMoonIllus.html or in *Planetarian* 14 (4) (Dec. 1985).

think the correct one was identified by science educator Carl Wenning in 1985. He noted that we perceive the sky to be farthest away from us in the direction of the horizon and closest to us directly overhead. It isn't, of course, but our brains somehow come to that conclusion. Looking toward the horizon, we see smaller objects on Earth as farther away than larger ones because we expect that the smaller things look, the farther away they must be. This, and perhaps other factors, apparently leads us to conclude that the sky "beyond" the Earth at the horizon is farther than the sky directly overhead, with no reference points between us and it.

The Moon really makes the same size image in our eyes throughout the night. Our minds compare that image to how far away the background seems to be and thereby determine how big we *think* the Moon is. *Against the apparently more distant sky at the horizon, we perceive the Moon to be larger than against the apparently closer sky overhead.* Even though you have done an experiment to see that the Moon doesn't change "size" throughout the night and I have presented a perfectly plausible perception-based explanation for this phenomenon, I'll bet you a nickel that the next time you see the Moon near the horizon, you will still think it especially large.

We Are Not Computers

Throughout our lives we memorize a stunning amount of information. Brain researchers of many stripes are hard at work trying to understand how we store, retrieve, process, and act upon it. Just as astronomers develop theories about how stars work or how the universe formed, brain researchers make theories about how the brain works. They can then test their theories with experiments using an array of modern technologies, such as MRI (nuclear magnetic resonance imaging), fMRI (functional magnetic resonance imaging), and PET (positron emission tomography). Functional MRI in particular can make high-resolution, high-speed images of brain activity. As a result, detailed locations of various mental functions are now being identified.

A photograph stores all the visual information about the subject in one place. Furthermore, the various pieces of the image are stored in physically correct relationship to each other. Computers generally store images, sounds, and other data in similar ways, by scanning the information and keeping the bits in a sequence that can be used to reconstruct a facsimile of the original data. However, as noted earlier, information stored in our brains is spread out in different regions. A single scene, for example, is separated into sights, sounds, smells, and other sensory data. Then these are broken up further and stored in various places. Different shapes go into different regions, different smells go into different regions, and so on. When needed, they are brought together and used to reconstruct the memory.

The process of creating permanent memories amounts to physically changing the neural connections in our brains. Once information is stored in long-term memory, it can be drawn upon when we think. Furthermore, because of the way memories are stored, they can be evoked completely out of the context in which they originated. For example, suppose you have recently decided that your favorite kind of apples is Royal Gala. You've memorized the taste, smell, color, and name and moved on with your life. Several months later, you are feeling a little ill. Walking down the street, you see a sign with a big red ball. For no reason you can think of, you recall the childhood adage, "an apple a day keeps the doctor away," figure, why not, it can't hurt, and decide to have an apple when you get home. Then you remember that you only have Red Delicious apples at home. Good enough? You remember the smell and taste of the different types of apple. No. You want a Royal Gala. So you arrange to pass a produce stand (or a grocery store if you don't live in a big city with produce stands) and buy a Royal Gala.

In this scenario, you have evaluated a concept, Royal Gala apples, deciding that you like them. You have then committed the name, flavor, taste, color, and your feeling about them to memory. Furthermore, from childhood you have developed ideas about apples. You know the shapes, colors, textures, and tastes of a variety of other kinds. You have also made connections between apples and other things in your life, such as the time your grandfather lifted you up to pick one

151 | Breaking Up Is Hard to Do

from a tree or the caramel-coated apple your mother bought you once as a reward for getting a perfect score on a spelling test, and the old adage about apples and doctors. Clearly, the information about apples, and Royal Galas in particular, didn't remain isolated in your mind. You used your knowledge of apples in your thinking and decision-making processes. In this situation, seeing the red ball reminded you about apples. You connected your feeling ill to your memories of their alleged medical benefits. Your ego directed you to desire your favorite apple, which brought to mind Royal Galas.

This kind of interconnectivity between memories and other aspects of our thinking is key to understanding why it is so hard for us to change our beliefs, especially concepts that we use in a variety of different contexts, such as associating the name Royal Gala with a taste we especially like. Once we have convictions like "Royal Galas are my favorite apples," or, in astronomy, that the Moon affects human behaviors and activities, they are hard to replace. Our resistance to changing beliefs depends on several factors, including our emotional attachment to them; how much they affect other beliefs we hold; the strength of the evidence against our beliefs; and whether we have better ideas to replace them. Let's consider each of these in turn.

EMOTIONAL TIES TO BELIEFS

From childhood onward, we gild our beliefs, our possessions, and our experiences with emotions. Children have favorite toys, stuffed animals, books, and television shows. While adults might say, "That's a really good book," a child will say, "I love that book." Even as adults, we have emotional attachments to inanimate objects. If we have a good time on a trip, we are much more likely to remember the details about it than if the trip was uneventful or unenjoyable. Many couples have special songs that played when they first met or at some other special time; these songs will always evoke memories and strong emotions for them.

Changing beliefs to which we have such strong emotional ties is very hard. Let's first consider changing a belief about something that carries little emotional weight, like your favorite type of apple. "Tastes" in things are emotional in that they have little or no rational basis. You cannot be persuaded by reason that one kind of apple tastes better than another. You can be told that one apple is sweeter or tarter than another, but in the end you have to taste them; only then can you decide which you like best.

"But lots of people read reviews of things like wine or cars and are influenced by the reviews."

That is quite true, of course, at least until the consumers try the product being recommended. An especially effective way to discover your own tastes is to do a blind test of a variety of similar things, such as different wines. That is, have someone serve you a few wines without letting you see the names of any of them.[3] Under those circumstances, you will be much less likely to agree with "experts" than if you know beforehand the names and reviews of the wines you are tasting.

In general, the opinions of others become less important after we experience the thing for ourselves. People who are more impressionable are more likely to accept the opinions longer, but personal tastes and experiences eventually overcome most other forces. Of course, sometimes we discover our true beliefs too late. "How in the world could I have believed that advertisement? These potato chips are terrible!"

What might make you change your mind about your favorite apple? On the positive side, you might try another type of apple and decide that you like its taste better. On the negative side, you might get bored with Royal Galas or you might hear from someone that Royal Galas have been shown to cause some illness (they haven't). If you trust the source of the news, your concern about illness may persuade you to find another type of apple to enjoy. The point is, something significant has to happen to convince you to change. It may be sudden or gradual, but you have to pass a threshold of resistance before you change.

---

[3] Even better, have the wines set out by someone who leaves the room before you come in, or make sure your server doesn't know which wines are which. That way, the server's expression can't bias you. This is the basis of a scientific double-blind experiment.

EFFECTS OF NEW IDEAS ON BELIEFS WE ALREADY HOLD

Now we turn from emotional associations to intellectual effects of new ideas. All of us have memorized facts for exams and then forgotten most of what we learned shortly after the exams were over. For the most part, this happens with subjects about which we know little, about which we have little interest, or about which we have strong beliefs that we don't plan to change even in light of new information. In the parlance of inventory control, these are situations of last in, first out for our memories.

Sometimes we memorize information, but it has no connection to anything else we know, so it remains isolated in our minds. For example, you may well have learned that Saturn has the most known moons, Jupiter has the second most, and Uranus has the third most. This was indeed true until 1997, when two moons orbiting Uranus were first discovered. Now Uranus has more known moons than Jupiter. This new piece of information has no bearing at all on anything else you think about. Therefore, you don't have to adjust any other beliefs to accept it.

Much of what we learn, however, is not held in such splendid isolation in our minds. We use it when we construct new beliefs or reinforce old ones. For example, if you believe that there was (or is) life on Mars, you will be comforted to learn that astronomers have found what appear to be dry riverbeds on the red planet. Since water is essential for the existence of life as we know it, evidence that water once flowed on Mars's surface would reinforce your belief.

Ideas that we have used in other contexts are especially hard to replace because they and their implications are interwoven in a variety of places in our brains. Rejecting a belief about something that we have used in other contexts requires us to rethink some or all of the implications of that belief. This is often extremely disconcerting because of the mind's interconnections; tracking·all the consequences of a belief is difficult because they are so complex. Following up on the life on Mars example, suppose you later learned that geologists had reevaluated these alleged dry riverbeds and concluded that they were all created by mechanisms other than water erosion, perhaps seismic activity, lava tube collapse, or tectonic plate motion. The first thing you would do with this new information is assess its damage to your belief system about life on Mars: "So they haven't found signs of water there. Yet. Big deal." This is clearly not a fatal blow to the belief that life existed on Mars, and so you proceed to buttress your original belief by rethinking the original and new water discovery claims.

You might begin by putting the new discovery in doubt: "If they changed their minds once, they can change them again." Or, "How do they know those riverlike features weren't made by water? Perhaps water doesn't flow the same way on Mars as it does on Earth. After all, Mars has less gravity, less air pressure, different average temperatures, and different surface materials than Earth. Do we really know how these things affect water flow? Closer images could show new evidence of water flow that we can't see yet."

Then you might embellish your current belief with alternate locations for water, just in case the geologists are correct: "Maybe there is liquid water under Mars's surface. This could easily be the case, since comets are rich in water and are known to have collided with objects in the solar system. Indeed, there is evidence of water from comets at our Moon's poles. So maybe a water-rich comet fell onto Mars, made a crater, and sank to a depth where heat from inside Mars melted it. Mars might be warm enough to have underground lakes or oceans even today. The same heat could enable life to form and evolve."[4] By the time you finish thinking about this issue, you probably will have refined your belief (water underground, rather than on the surface) but not changed it.

---

[4] I personally believe that primitive life forms will be discovered on or inside Mars. Also, riverbedlike features on Mars definitely appear to have been created by water flow. I use the alternative origins of these regions just to make the point that we tend to accept ideas or results in accord with our beliefs and reject or devalue those with which we disagree. Using a moderate-power telescope, you can see curvy riverbedlike features on our Moon too. These were made when underground rivers of magma solidified. The solid magma shrank, leaving gaps above it. Then the ceilings of these voids collapsed, leaving what we call rilles on the lunar surface.

THE STRENGTH OF THE EVIDENCE AGAINST OUR BELIEFS
AND SUITABLE REPLACEMENTS

The more we use a belief in developing other beliefs, the more we will work to maintain it in the face of contradictory evidence. However, there can come a time when the weight of evidence against a cherished idea is overwhelming. In such an instance, we not only have to replace the original belief but also have to tear apart other concepts we constructed with it and replace them.

As an extreme example of adapting to a change in beliefs or perspectives, consider what we each have to do to deal with the death of someone we love. Our view of the world includes that person's presence and his or her responses to us. The pain of not being able to see that person, touch them, talk to them, hug them, spend time with them, and know they care about us is initially unbearable. Therefore, one of our first responses may be to reject the new reality. Accepting it requires too much adjustment, too much rewiring of mental connections as our brains literally reconstruct our perception of the world. You go near the house where that person used to live and anticipate seeing them, but they don't greet you at the door. You go inside and expect to see them sitting in a favorite chair, but they aren't there. Trying to reject such a change but eventually being forced to face its reality is an emotional roller coaster unlike any other. Both your emotional attachment to the person and their deep connections to other parts of your life make accepting their death incomparably difficult.

Breaking Up Is Hard to Do

Let's consider a sequence of astronomical beliefs that caused great anguish for different generations of astronomers. The issue is the size and fate of the universe. We begin by revisiting the Earth-centered cosmology.

Observations led early Babylonian and Greek astronomers, among others, to identify eight different types of permanent objects in space: the Sun, the Moon, five planets (Mercury, Venus, Mars, Jupiter, and

Saturn), and the "fixed" stars. Observations of the daily motion of the Sun and Moon and nightly motion of the stars and planets led to the common-sense belief that they all orbit the Earth.

Early proposals by Philolaus and Aristarchus that the Earth orbits the Sun met with common-sense objections. If the Earth were moving, wouldn't we feel it? If the Earth were moving, wouldn't the planet rush through the air, creating enormous perpetual winds? Without reasonable answers to such questions, the consensus in ancient times was that the Earth was fixed and everything orbited it. Aristotle, Hipparchus, and Ptolemy crystallized this belief in a geocentric model of the universe.

Furthermore, common sense dictated to ancient philosophers that the orbits of the heavenly bodies around the Earth should be circular. This followed from the belief that circles were the most "perfect" shapes and that everything cosmic would "necessarily" follow the most perfect motion. "Perfect" and "necessarily" are among a variety of terms frequently used without justification or definition by early philosophers when they wrote about the natural world.

To accommodate such phenomena as the occasional reverse (retrograde) motion of the planets among the stars, Ptolemy's version of the geocentric model of the universe contained circles within circles. While a planet's bigger circle (its deferent, or orbital path) always moved in the same direction around the Earth, the motion of its smaller circle (its epicycle, or rotation around the deferent) would occasionally make the planet appear to be moving backward in the sky (see figure 5.1). As better observations were made, this theory had to be modified with epicycles on epicycles and off-center circular motion to more accurately represent the movement of these bodies. Though the system became incredibly cumbersome, the appeal of the geocentric model and the authority of "the ancients" were so strong that it held sway for 1,400 years. Interestingly, in proposing the heliocentric (Sun-centered) universe in 1543, Copernicus was trying to recover a simple cosmic system based on circular motion. As noted in chapter 2, Copernicus's Sun-centered system gave planet positions no more accurate than those in the by then highly refined geocentric model.

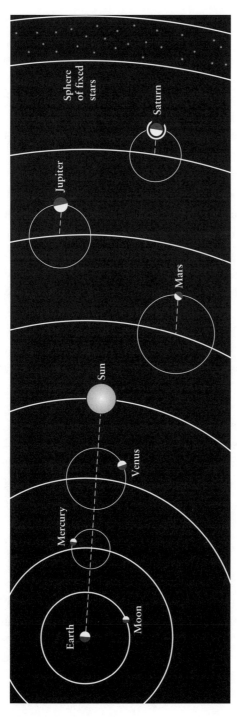

FIGURE 5.1 The geocentric model of the universe. All the celestial bodies were believed to orbit the Earth with circular orbits. Nonuniform motions of some celestial bodies were explained via epicycles: small circular paths centered on those bodies' orbits around the Earth.

This brings us to the first revolution of this section. Initially, Copernicus's belief in a Sun-centered universe had few adherents outside astronomical circles. The fact that the Copernican system made predictions of planet locations that were no better than those of the geocentric model didn't help. Unless it could be shown to be more accurate, the Copernican system could be ignored by believers in the geocentric model as just "another" model and not the one given authority by the "ancients."

Something else happened to undermine the idea of a geocentric universe even before the Copernican model was corrected. As mentioned above, one of the basic tenets of the geocentric model was that all objects in space orbit the Earth. In his 1610 observations, Galileo shattered that belief. With his crude telescope, Galileo observed three "stars" near Jupiter. Over a period of a week, he also discovered a fourth "star" and noted that all four new bodies remained in Jupiter's vicinity. This made them different than the other stars, through whose regions Jupiter moved during that time. He (correctly) concluded that the new bodies were actually moons orbiting Jupiter. Galileo had demonstrated that objects exist in the universe that do not orbit the Earth. This opened the door for the younger generation of astronomers, who were not in thrall to the ancient geocentric theory, to delve deeper into the heliocentric model and make it work.

Galileo published this discovery in March 1610 in a book, *Sidereus Nuncius* (Starry Messenger). The Catholic Church had long subscribed to the geocentric model, which supported the belief that God created the Earth as a special, unique place. The geocentric model also fit with the need of the normal human ego to be at the center of everything, regardless of religion. Galileo's observations delivered a stunning blow.

The discovery of Jupiter's moons was doubted or ridiculed by many people, especially those who did not have telescopes to see for themselves. The book's publication also led to the beginning of Galileo's troubles with the Inquisition, which would dog him for the rest of his life. *Sidereus Nuncius* was listed in the Roman Catholic Church's Index of Banned Books from 1616 to 1822, which limited its readership, at least in the early days. In the same year, 1616, Copernicus's earlier work, *De Revolutionibus Orbium Coelestium* (On the Revolutions of the Heavenly Spheres), was also put in the Index.

In 1611, Johannes Kepler confirmed Galileo's discoveries of Jupiter's four largest moons and dealt another undermining blow to the geocentric model by identifying a better one. Kepler began with Copernicus's heliocentric model and (correctly) decided that the error Copernicus had made was assuming that the planets moved in circular orbits. Kepler believed that with sufficiently accurate observations, he could determine the true shape of planetary orbits. He obtained these precise observations from his mentor, astronomer Tycho Brahe, and after many attempts, some of them driven by mystic beliefs, he found a geometric shape that was consistent with the orbital data. Between 1609 and 1619 Kepler worked out three laws of motion based on a heliocentric model with the planets moving in elliptical orbits. The strikingly accurate predictions of this heliocentric model left no doubt in his mind that it was the better description of the solar system. The problem lay in convincing other people.

Yes, But Why Ellipses?

The change to complete belief in elliptical orbits was a rocky one. Few of the older generation of laypeople or astronomers were willing to let go of the ancient view of circular orbits; the prevailing belief was perfect to explain life in a perfect universe. It was primarily the younger generation who were swayed by the predictive power of Kepler's laws rather than the historical power of antiquity. Yet a great problem remained in Kepler's version of the heliocentric universe: there was no explanation *why* the orbits of planets around the Sun or moons around the planets were (are) elliptical.

The first step toward discovering that explanation was taken in 1687 when Isaac Newton published his *Principia Mathematica*, which sets out both his laws of motion and his law of gravitation. The latter is a mathematical equation that predicts how strongly pairs of objects in the universe attract each other. One of its predictions is that all objects attract each other. This had cosmic and theological implications. It meant that all the stars are pulling on each other, even if they appear unmoving. Furthermore, they didn't seem to be uniformly distributed. Newton calculated that fixed, randomly distributed stars should eventually fall together due to their mutual gravitational attraction. These larger clumps of matter would then pull other stars into them until the universe coalesced and ceased to exist.

The implications greatly disturbed Newton. If the stars visible to the naked eye comprised the entire universe, then it would take a mighty act of God to hold everything in place. But Newton didn't need to resort to this construct. He found a more palatable solution using the telescope, which revealed myriad stars that are too dim to be seen with the naked eye. In an effort to prevent the universe from collapsing, Newton deduced (not entirely correctly) that all these dimmer stars were farther away. Generalizing from all the dim stars he saw, he proposed that the universe was infinite in extent and that stars are distributed evenly throughout it.

The assumption of an infinite universe and a smooth distribution of stars provided Newton with an explanation of how the stars, and therefore the universe as a whole, could remain unmoving and therefore last forever. His law of gravitation predicted that the forces of gravity among the stars in a perfectly smooth and infinite distribution would cancel one another out, leaving all the stars fixed in their places. Newton did realize that this situation was highly unstable. The slightest movement of even one star in space would cause the entire system to eventually collapse together. He believed that one of God's roles was to prevent that from happening.

The concept of an infinite universe raised perplexing questions for the Church. A crucial one was: Where is heaven located? In the Ptolemaic system of crystal spheres, heaven was believed to exist beyond the farthest sphere containing the stars. Making the transition to belief in an infinite universe caused considerable chaos, which may have contributed to the burning of one of the theory's early (pre-Newtonian) advocates, Giordano Bruno, by the Church in 1600. The idea was rejected by most people until Newton provided a compelling reason to believe it. The eventual acceptance of Newton's infinite, static universe set the stage for yet another traumatic transition in belief about the universe that occurred much later—in the twentieth century.

THE ACCEPTANCE OF ERRORS

Experiments showed that Newton's law of gravitation worked, but Newton never claimed to know why. The next step in understanding the gravitational attraction among all the objects in the universe came in 1915, when physicist Albert Einstein published his general theory of relativity. Among other things, it provided a much deeper explanation of how gravity works than Newton's law.[5] Furthermore, in its simplest form, general relativity predicts that the universe should be either expanding or contracting. Influenced by the then-prevailing Newtonian belief in the static universe, Einstein used a more complex version of the equations that accommodated that notion. He did this by adding a term that represents a force counterbalancing the need for expansion or contraction.

In 1929, Edwin Hubble published observations showing that the universe is expanding. He had observed distant galaxies and discovered that all of their light shifted in color. The farther away they were, the more their color was shifted. Furthermore, all the distant galaxies' colors were shifted toward the red end of the light spectrum. Hubble interpreted this red shift as the result of all those galaxies moving away from Earth.[6] The red shift is analogous to the deepening in pitch you hear when a whistle or siren moves away from you. This is called a Doppler shift, after Christian Doppler, who explained it in 1842. By contrast, the analogue to the increase in pitch of a siren moving toward you is a color shift toward blue (blue shift).

Faced with irrefutable evidence that one of the building blocks of his static universe model was wrong, Einstein and others removed the extra term, called the cosmological constant. Since then, astronomers have used the equations of general relativity in the form that predicts an expanding universe. Einstein called the cosmological constant "the

---

[5] Discussing the details of general relativity would take us beyond the scope of this book, but you can learn more about it in Kip Thorne's *Black Holes and Time Warps*, among other books.

[6] Contrary to common sense, this does *not* mean that we are at the center of the universe. If you make the same observations from any other galaxy, you get the same results!

biggest blunder of my career." Because he chose to remain within the Newtonian framework for the universe, Einstein missed his opportunity to predict that the universe expands.[7]

Hubble's discovery caused considerable rethinking about science and theology throughout the twentieth century. Irrefutable evidence to the contrary combined with a new theory had forced out the geocentric universe model in the seventeenth century. In that same century, Newton made a compelling case for a static, infinite universe. By 1929, there was nearly irrefutable evidence that Newton's static universe was invalid. Old theories and cherished beliefs die hard. Counterexplanations were therefore put forward to explain the apparent expansion of the universe. Paramount among these were a group of ideas called tired photon theories. They proposed a nonexpanding universe in which the light particles, or photons, lost energy by a variety of mechanisms as they traveled through space toward us. When a photon loses energy, it grows redder. The origins of red shifts in tired photon models ranged from collisions between light and intergalactic matter to changes in the laws of physics among the distant galaxies around us. None of these theories has stood up to observational tests.

The Big Bang

A replacement theory for Newton's static universe was needed. Put yourself in the shoes of someone early in the twentieth century who had been raised to believe, as most people did, that the universe was both infinite and static. Then in 1929, you learned suddenly of compelling evidence that the universe is actually expanding. Allowing that you accepted the observations and analyses, how would you have incorporated this new concept into your view of the universe?

---

[7] It is very interesting to note that in 1998, observations revealed that the universe is expanding outward faster than can be explained just by the force of the Big Bang explosion, from which most astronomers believe it formed. In fact, it is apparently *accelerating* outward, which would require an ongoing outward (antigravity) force. To account for this phenomenon, astronomers are returning to the rejected cosmological constant term, putting it back into the equations.

If you were like most people, you would find the least change necessary to accommodate the new information. Specifically, you would have to allow that the universe is expanding (not static), but perhaps you would keep the belief that it is infinite. Models were developed for an expanding infinite universe. According to these Steady State models, as the universe expands, new matter is created to replace the matter that has moved away. New physics is required to explain the creation of this matter but it would not be impossible to develop. Furthermore, the amount of matter that would have to be created would be below the minimum that we can observe today (that is, observations cannot yet rule out the formation of this new matter). The Steady State model would require only one new atom to form in every ten billion cubic meters of space each year.

Steady State theories were extremely popular throughout the first half of the last century for the simple reason that they did the least damage to people's prior beliefs. However, in 1964, engineers working for Bell Laboratories discovered that the entire universe is filled with microwave (short-wavelength radio wave) radiation. Furthermore, detailed study showed that this radiation was uniform throughout the cosmos at about one part per billion. Steady State theories had no explanation for this cosmic microwave background radiation. But another set of theories waiting in the wings predicted it precisely: Big Bang models.

If, as an early twentieth-century science buff, you also rejected your belief that the universe is infinite, you would have inferred from the expansion of the universe that it used to be smaller. Run time backward and the universe contracts—its volume decreases. Big Bang models predict that the universe was once extremely small—in everyday terms, smaller than a golf ball.[8] Again, prior beliefs would come into

---

[8] This analogy is actually misleading because the expansion of the matter and energy in the universe coincides with the expansion of space itself. Therefore, when the universe was smaller, so was the space in which all its matter and energy resided. When it was the size of a golf ball, that was also the volume of space. There was no separate "outside" in which to exist and measure the tiny, expanding universe.

play. Since the universe used to be much smaller, you would have to assume that it began at some finite time in the past. One of the prior beliefs common in the early part of the twentieth century was that the universe was eternal. To eliminate this belief too was quite intolerable to many people since it brought into question the purpose of life, an essential component of most religions.

Therefore, Big Bang models were developed that would allow for the universe to have existed forever. They assert that we live in an oscillating universe that expands to a maximum size and stops under the influence of the gravitational attraction of all its matter pulling on all the other matter. The universe then recollapses to a very small volume and then violently reexpands. The idea of an oscillating universe became popular after the Steady State theory was refuted because it swept the question of how the universe began under the rug, declaring that it has always been here, just changing form. This represented the minimum adjustment needed to make prior beliefs consistent with new observations.

The oscillating universe idea has recently been struck a perhaps fatal blow by observations that the universe "probably" will expand forever. I put "probably" in quotes because during my career, successive observations have supported the belief that the universe will contract again (necessary for an oscillating universe) and the belief that it will expand forever. This change in perspective has occurred at least a dozen times between 1980 and 2000. However, our technology for making observations and our understanding of astrophysics have advanced so much that most practicing astronomers now expect the latest observations, consistent with a permanently expanding universe, to be confirmed in the next few years.

Therefore, astronomers have growing confidence in the Big Bang theory that says the universe began as a tiny entity, has expanded and gone through a variety of changes, has developed the capability to support life, and will expand forever. Scientists don't yet know how the universe began. I expect that that will be discovered in the twenty-first century. However, starting from less than one second after the universe came into existence, the equations of the Big Bang model provide us with tremendous insight and accurate predictive power about how it has behaved since then. One of the lessons of this situation applies to virtually all science: even though scientific theories are incomplete, they provide important understanding of how various parts of nature work.

It is sometimes easier to reject the facts than to accept something alien to your belief system. Because the single Big Bang theory has taken us so far from common-sense models of the universe, it still presents great difficulty for many people. Indeed, some scientists and others still believe in the Steady State theory or the oscillating Big Bang theory, in spite of the observational evidence against them. Whether for theological, philosophical, or aesthetic reasons, many people choose to believe that the universe is infinite and eternal, or at least relatively unchanging. For example, the Steady State model allows for life to continue forever into the future. The future of a universe that expands forever from a single Big Bang is certainly more bleak. That model predicts that eventually all the stars will expend their fuel and extinguish themselves, thereby ending the existence of life as we know it.

A Personal Cosmology

Let me end this chapter by recounting how personal experience has taught me that change in beliefs can happen, even in matters as big as the universe. As a child growing up in the 1950s and '60s, I developed a personal cosmology. I believed that we humans are truly making progress in understanding how the universe operates and that eventually we will know all there is to know. Granted, that could take millions of years, but there didn't seem to be anything to stop this advance, if we didn't blow ourselves up first. Armed with such knowledge, I expected that we would eventually be able to build more and more complex and sophisticated things and control more and more of our environment. Following this theory to its apparent conclusion, I came to believe that eventually we would not have anything new to learn or do. This rather depressing thought was compounded by the belief that if there were life after death, you'd eventually get pretty bored after a few billion years of cavorting.

The oscillating universe seemed to present a neat way out of all these conundrums. In the first place, we wouldn't have to worry about having nothing to do, either here or hereafter, because eventually everything in our present universe would be destroyed by collapsing together. The best part of this idea was that by the time of the "big crunch," as it is called, our descendants would be so advanced that they could take part in determining a new and better set of laws of physics for the next universe. I wondered whether our present laws of physics had been designed in part by life forms in the last incarnation of the universe.

You'll notice that my ideas about the future of the universe were derived from a deeply anthropocentric point of view—what is best for humans. That belief presupposes that the universe was made for us, which is a pretty egocentric, though not uncommon, belief. While the origins of my personal cosmology faded into recollections of my childhood, the resulting belief in the oscillating universe became fundamental to my view of the cosmos. I even carried it into my professional career. Every time observations suggested that the universe would expand forever, I felt confident that a few months later, results confirming the recollapse would be made, and they were. Then, in 1998, in the face of compelling observations that the universe will expand forever, I sat back to rethink my belief in the oscillating system.

The first thing I discovered was that my anthropocentric motivations for the universe were gone. I no longer believed that the universe was "made" so that we can be here. Furthermore, I no longer felt it necessary to worry about the job prospects of my descendants. It was a very liberating experience to realize that I didn't *need* to believe something (the oscillating universe theory) that was inconsistent with observations. Granted, I would have to change some other beliefs, but that seemed an adventure, going against the unacceptable ideas I had once held. Would they trip me up? Yes, from time to time they did and do, but at least I wouldn't have to try to make excuses for believing in something that would never happen.

This freedom of thought allowed me to consider something else that had been bothering me for a while, namely, eternity. The oscillating universe had allegedly existed forever. I get a headache from trying to figure out how anything could have been around that long. It seems to be the ultimate "something for nothing." Granted, we don't yet know how to explain the origin of a single Big Bang, but our scientific theories of nature are bringing us closer and closer to understanding that event. So I feel comfortable with the belief that some scientific mechanism for the creation of the universe will be discovered. A single Big Bang that expands forever no longer seems unacceptable. It still holds out the prospect of a barren future, but that is hundreds of billions or trillions of years down the road. Perhaps we will learn ways to enter different universes or create new ones. Who knows what discoveries will be made in that time?

6

The Sage on the Stage or the Guide by Your Side

A PEEK BEHIND THE EFFORT TO HELP YOU UNLEARN MISCONCEPTIONS

In the late 1970s I began teaching college astronomy. There I was, in front of 250 students for the first time, and I had never received a minute's instruction on how to teach. Back in those days, young college faculty were expected to have picked up the basics of teaching from the instructors they had had over the years. Of course, their professors had also learned to teach without any formal instruction in education, so it was mostly a case of the blind leading the blind.

As I can testify, the intuitive feeling a new and untrained teacher has about presenting material is that capable students should be able to comprehend clear explanations and make that knowledge part of their understanding about nature. This has been codified in the concept of the student as a blank slate (*tabula rasa*) ready to be filled with information. We now recognize how naive this idea is. How many courses in school or college did you take in which you went in on the first day with no beliefs at all about the subject?

It wasn't until late in the twentieth century that most educators "discovered" that they have to help students unlearn incorrect prior beliefs before they can retain and use correct information. Teacher training courses often emphasize this, but it really sinks home only after you teach a conventional class and see how many incorrect ideas students have at the beginning and which ones they retain afterward. For those of us who discovered the need to correct misconceptions on our own, the realization that students quickly forget much of what you normally teach came as quite a shock.

The reality is that most students come to most classes at all levels of their schooling with preconceived ideas about each subject. This happens because we all develop beliefs about most things we encounter, even if we have never had any formal education on them. The same argument applies to subjects we never learn about in school. I never took a course in economics, but I have a variety of ideas about it, and I'll bet you another nickel that many of them are wrong. Likewise, even if you have never taken an astronomy course, if you pick up a popular-level astronomy book, you will probably find that you already have beliefs about the material presented.

There is no magic pill to help people unlearn deep-seated incorrect beliefs (that is, misconceptions). If there were, we science teachers would be dispensing it like candy and many fewer erroneous ideas would be floating around. One thing that has grown clear to me in trying to help students replace misconceptions is that these beliefs must be addressed directly, intensively, and persistently if there is to be any hope of permanent change.

Two Roads to Rome

There are two basic ways to unlearn misconceptions about science. Both are based on the premise that in order to change, you must first identify your (incorrect) belief about something and then actively compare it to how nature really works. This process starts with a question, like "Why is the sky blue?" You might give the answer, "Because it contains a blue gas." The two methods of dealing with this incorrect belief differ in the way the teacher interacts with you when considering the validity of your idea. In one method, the teacher provides gentle, positive guidance as you figure out an experiment to test your belief, find the results of the experiment to be inconsistent with your idea, and construct a correct belief. In the other, the teacher aggressively confronts your incorrect belief every step of the way with irrefutable arguments and experiments.

These methods can lead to the replacement of deep-seated incorrect beliefs with correct beliefs, a process called conceptual change, but to appropriate a refrain from a Beatles tune, "you know it don't come easy." Some people respond well to the supportive approach, others to the confrontational. The supportive method has the disadvantage that it takes a lot of time and effort on both the student's and the teacher's parts. It takes much more time to help students "discover" important concepts than to tell them the same things. Discovery-based learning is effective in leading to conceptual change, but the number of concepts that can be discovered in a semester or a year is many fewer than the number that can be acquired by more conventional teaching—especially when combined with the more aggressive methods of addressing misconceptions. While more efficient, the confrontational method has the disadvantage of being, well, confrontational. I will have more to say about this issue shortly.

Different Strokes

Different people learn best by different means because of their diverse strengths and weaknesses. Some people have better memories than others; some people are better at comprehending what they learn; some are better at applying what they learn; some are better at analyzing what they learn; some have greater capacity to synthesize ideas from different realms; some are better at evaluating the information they have. Furthermore, some people are more visual, some are more verbal; some people are more receptive to information from authority figures than others; some are more organized in their learning habits; some come to a topic with lots of prior beliefs; some have relatively few notions on the subject.

Unfortunately, in real life, the range of teaching methods is limited, and few if any are truly optimal for anyone. The discovery-type experiences just described are too expensive to be implemented everywhere. The traditional compromise that best supports discovery learning is classes with laboratories and homework.

It is pretty well established that a typical large lecture class, say 250 students in introductory astronomy, is among the least effective ways to cause conceptual change. The major reasons are varied. As you probably know, students respond to the personalities of their teachers. A boring or inarticulate teacher can make even the most interesting course drag. Also, teachers often teach at levels inappropriate to their students. Even if all instructors were stellar, intrinsic problems remain. They boil down to the fact that if students aren't actively involved in the learning process, they are exceedingly unlikely to retain information, incorporate it into their understanding of the subject, or replace their incorrect beliefs with more accurate ones. Typically, the individual in a group of several hundred is unlikely to get actively involved in class discussion. But such classes are necessary evils, given the numbers of students and the numbers of teachers, classrooms, and pieces of essential equipment available.

"But you teach big classes just like you describe. How can you live with yourself if you know your students aren't really learning what you teach?"

Once I realized that normal lecturing is relatively ineffective, I couldn't teach that way anymore. I still had the responsibility of such large classes, but I was determined to make the learning experience more effective. First, I needed to know what incorrect beliefs the students had coming into class. Then I needed to find ways of making them realize that these beliefs were wrong. And finally, I needed to find effective ways of presenting the correct information. This process took a decade. I would never claim that the result is a course as effective in teaching as small, hands-on classes are. Nor is it right for all my students. Nor did I see ahead of time all the twists and turns in the process. But eventually I did arrive at an approach that makes the knowledge presented in a large class compelling. And it all began innocently enough.

SHOW, DON'T TELL

In order to help people deal with their misconceptions, we have to know what they are. There are at least two ways of getting this information: asking and taking. Asking for the information requires that students actively compare what they are learning with what they believe. Armed with insights from small discussion groups I held with students about their misconceptions, I realized that there are so many incorrect beliefs (both superficial and deep seated) that I would have to spend centuries interviewing a few people at a time to get at many of them. Or I could enlist thousands of volunteers and get most of the information in a few years. I chose the latter route and began the very next semester.

During the first class of each semester since then, I have begun with a demonstration designed to convince everyone that they have incorrect beliefs about science. I put an electric air gun, like the ones used to blow leaves off a driveway, on the table. The gun is fixed in a base that points it straight upward. I then produce a beach ball and throw it out for the students to bounce around for a while. Then I ask three questions:

1. What will happen if I turn the air on and throw the ball up into the vertical airstream?

About half the students have seen this demonstration and respond, correctly, that the ball will hover above the air gun. The other half expect that it will fly away. I then do the demonstration and, indeed, the ball hovers several feet above the air gun. Over the roar of the air, I shout the second question:

2. What will happen if I tilt the air gun over toward them at an angle of, say, 45 degrees?

Virtually everyone expects that the ball will be blown into the audience. I do this experiment, and the ball moves sideways and hovers over nothing about five feet to the side of the air gun. I then slowly rotate the gun around and the ball follows in a broad circle.

Finally, I ask: 3. What will happen when I turn the air gun off?

There is universal, if slightly more hesitant, assertion that the ball will drop. It doesn't, of course. As the air stream decreases in strength, the ball moves back to the mouth of the air gun, where I catch it (I've practiced).

This demonstration is truly effective in getting the students' attention. I go on to explain that they will be learning many things that contradict what they already believe. I give a few astronomical examples of common incorrect beliefs, such as that the cause of the seasons is Earth's changing distance to the Sun; that depleted ozone in the atmosphere is not replaced; that the primary purpose of a telescope is to magnify; and that Mercury is the hottest planet. Then I offer an incentive of extra credit to those students who provide me with a list at the end of the semester of 42 astronomical beliefs that I have corrected for them. (There are 42 contact hours in the semester.) Each item has to include a sentence stating what their original, incorrect belief was and where it came from, if they remember. This is how I get them to volunteer their incorrect beliefs.

Of course, there is a lot of duplication in the 50,000 statements submitted over the years. By combining identical items, I began developing a list of incorrect beliefs, which now number more than 1,560. As noted in chapter 1, you can find a complete list of them at http://www.umephy.maine.edu/ncomins/.

Some statements are given quite frequently, such as "I thought that Polaris is the brightest star" and "I thought the asteroids were close together, like in the movies." These are clearly common incorrect beliefs. However, even many of the less frequently cited statements are quite common. This becomes apparent when I ask questions on tests based on some of these less common beliefs. For example, only a few students have ever stated in their lists that they thought stars have molten liquid surfaces. I suspect, however, that this is a common belief; even after I have taught the correct science (gas all the way down), a large plurality of students choose the answer on an exam that stellar surfaces are liquid. This latter process of getting information by asking directed questions is an example of "taking" it. I will give another example of taking information shortly.

"It's not fair to ask loaded questions like that on tests."

I don't ask loaded questions as a common practice, but they sometimes are instructive in showing just how deep-seated some beliefs are. Also, keep in mind that I had already taught the correct answer. The fact that many people didn't "get it" was as much my fault as theirs. If I had done a better job teaching by helping them confront their prior, erroneous belief, more of them would have gotten the right answer.

BOOK LEARNING

Armed with lots of incorrect beliefs, I began reading the literature in search of ways to help my students in a large lecture setting. The results were disappointing. Most success in correcting wrong ideas has come in small classroom settings, especially, as noted earlier, when there are many resources available for students to use in their process of constructing new beliefs. One day, late in 1994, while reading the journal *Physics Today*, I came across an obituary for the astronomy writer William J. Kaufmann III. I was shocked and saddened. He was only in his mid-fifties at the time and was writing some very successful college astronomy textbooks, including one right at the level of the introductory course I taught. In an epiphany, I realized that here was an opportunity to help students in classes of all sizes address their misconceptions by writing about them in a textbook. Perhaps I could figure out a way to get students to think about their preconceptions before reading the correct science, then guide them through the reasoning that would help them unravel their incorrect ideas while making the correct ideas compelling.

The first big obstacle was getting the right to rewrite Kaufmann's book. I called the publishing house and talked to the editor in charge of acquiring astronomy texts and authors. She was a very pleasant, very sharp person. After sharing our sorrow at Bill's passing, I pitched myself to take over his paperback text, which I knew from the copyright date was due to be rewritten soon. I sent her copies of my previous book and trade articles. She then asked me to write some sample chapters for the revised text. While writing these, I worked out a way of addressing and correcting incorrect beliefs. I sent the sample chapters off. Several weeks later, I got a phone call: "Neil, we'd like you to take over the book, but there is a real time crunch. We need a first draft in six weeks." They bought me a shiny new computer (66 MHz with an Intel 486 CPU, no less!), and six weeks later I shipped the first draft to New York.

There was deafening silence from the publisher for several weeks. Then my editor called, saying that while the draft would need some work, it was acceptable and that the misconception-based ideas seemed very promising. Then she told me that a normal revision of that text took eighteen months, so I didn't feel bad that my first draft "needed some work." The book came out on schedule, and the idea of addressing misconceptions in a text was quickly accepted by my teaching colleagues around the world.

My approach to misconceptions in the textbook has several parts. At the beginning of each chapter I ask a series of questions based on common misconceptions about the material in that chapter. The idea, of course, is to have readers thinking about their ideas as they read the material. When approaching the section of the chapter in which the correct information is presented, I try where possible to explain why some of the common incorrect ideas are wrong. A numbered icon appears next to each paragraph containing the correct explanation for each question. At the end of the chapter, the questions are repeated with brief summaries of the correct explanations given directly below them.

Student response to this feature has for the most part been positive. The only exceptions I have heard about are people who have difficulty with the fact that someone else seems to know what they are thinking. In other words, they feel uncomfortable being wrong and knowing that someone else *knows* they are wrong. This became particularly clear to me after a talk I gave about common misconceptions taken right from the book. A student came up to me afterward looking shell shocked. "I can't believe it. You were describing me exactly," she said. "I believed all of those things." I assured her she wasn't alone and that those incorrect beliefs didn't mean she was stupid. "We all have lots of misconceptions. They're unavoidable." I don't know whether or not she found that reassuring.

IN YOUR FACE

Having my own textbook that addresses common misconceptions proved a valuable first step in changing how I taught large classes. But addressing incorrect beliefs in a text is one thing; dealing with them in the classroom in order to help students retain correct knowledge requires active teacher intervention. I wish I could say that I planned what happened next in that regard, but it came unexpectedly as a side effect of the need to take class attendance. I decided to change my *laissez-faire* policy in light of reports that improved attendance leads to better grades among some groups of students. I made showing up to class worth 10 percent of the grade. This straightforward decision set into motion a chain of events that dramatically changed the way I teach.

The reality of college student life is that there are classes you have to take but choose not to attend very often. Sometimes you can learn the material from the book and don't want to waste your time going to class. Sometimes the professor is deadly boring. Sometimes you have a hangover. Sometimes it's too early in the morning. Sometimes you just hate large classes or discussion groups, or something else entirely. Nevertheless, many educators' experiences reveal that attendance improves many students' grades. The question for me was how to take attendance, knowing that some students will do anything possible not to attend and still get credit for being there. You can't just have students check off their name on a list or even sign in personally; they quickly learn that they can sign in for their friends, and attendance drops. So I came up with the idea of asking the students a question at the end of each class and requiring them to write it down, write an answer, and hand it in on a sheet of paper. Since I thought up the question as each class ended, the students would have to be there to know it and answer it.

In light of this book's subject, you won't be surprised that I decided to ask questions based on common misconceptions. Furthermore, they were on topics about which I had not yet presented any information in class, so the answers represented what students thought previously. This is the second way of "taking" information. I assured the students that during the semester I would only scan their answers to be sure they were responding to the question I asked (and therefore had been in class). I also promised them that their answers would have no bearing on their grades and that I would never use their names in connection with any of their answers. I didn't forbid discussing the answer while they were writing, but I didn't encourage it. I answered each question at the beginning of the next lecture. Here are some of the questions I asked in the fall of 1998.

1. How did the Moon form?
2. a) How much of the Moon's surface can we see from Earth over the course of a year? b) Does the Moon rotate?
3. What causes the seasons?
4. What causes the tides?
5. What causes the phases of the Moon?
6. What is a blue moon?
7. What is the shape of the Earth's orbit around the Sun?
8. How many zodiac constellations are there?
9. On which planet is the surface temperature highest?
10. What is the origin of the "face" on Mars?
11. Describe the rings of Saturn.
12. What fraction of the solar system's mass is in the Sun?
13. Why does the Sun shine?
14. How does the Sun compare in size to other stars?
15. What is the distance between the solar system and the star closest to it?
16. Which stars "last" longer, higher-mass stars or lower-mass stars?
17. How many stars are there in the solar system?
18. How many stars are there in the Milky Way galaxy?
19. How many galaxies are there in the universe?
20. If you go out at noon today, when the Sun is nearly highest in the sky, at which of the following angles above the southern horizon will you find it? Choose the closest angle: 30°, 45°, 60°, 90°.
21. How did the universe begin?
22. Please evaluate these questions. Have they had any bearing on how you think?

At first, many students were angry at having to write down their beliefs only to learn at the beginning of the next lecture that they were often wrong. This I had expected. What I hadn't anticipated was the evolution of their attitude toward this process. After two or three weeks, I began to see a noticeable change as they answered the questions and turned them in. Their resentment gradually decreased, and most students appeared to tolerate the questions. By the fourth week there was an air of anticipation for each question. I began seeing groups of students discussing how they had answered the day's question as they were leaving the classroom.

By the middle of the semester, the majority of the students were actively involved in the process of addressing their prior beliefs, driven by these questions. I mark the beginning of this period, which ran through the end of the semester, as the first time that loud cheering and moaning occurred in response to the correct answer I presented. I observed students who got it right raising their arms as they would when their team gets a touchdown or giving each other high fives. Students began asking me questions about why their own ideas were incorrect. The discussions among themselves were noticeably more intense. After each class, there were perhaps half a dozen groups of students standing around talking about the questions, arguing for one belief or another, even after I left. Several students told me that they went home after each class to research the question in the textbook or on the Web. Others told me that they would ask their families the questions when they went home for visits.

This attendance tool had morphed into two other things: a way for me to collect data on the percentage of students who held incorrect prior beliefs and a way to get students actively involved in thinking about their prior beliefs.

As the semester unfolded, the power of the attendance questions became even more apparent. This also gave me the opportunity to see if I had really changed anybody's misconceptions. I decided to administer a nongraded test near the end of the course asking the same questions as had been given throughout the semester. I gave this test in the penultimate week of classes, before anyone had started studying for the final exam. The idea was to see if after several, or even a dozen, weeks with the new concepts, students had changed their beliefs. Some typical results from the 1999 class follow.

Cause of the Seasons
Asked on 9/7/99

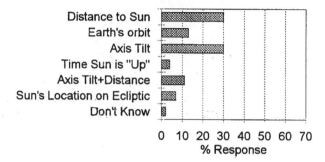

Precentage of Solar System Mass in Sun
Asked on 9/28/99

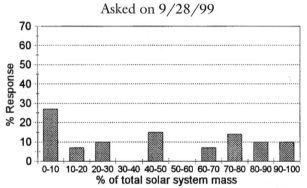

Which Last Longer: Low- or Hi-Mass Stars
Asked on 11/9/99

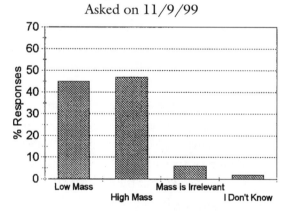

FIGURE 6.1 Results of misconception-based attendance questions—seasons, Sun, stars.

Cause of the Seasons
Asked on 12/16/99

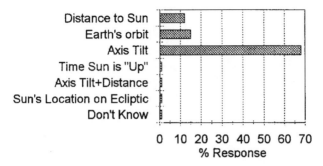

	0 10 20 30 40 50 60 70
Distance to Sun	
Earth's orbit	
Axis Tilt	
Time Sun is "Up"	
Axis Tilt+Distance	
Sun's Location on Ecliptic	
Don't Know	

% Response

Percentage of Solar System Mass in Sun
Asked on 12/16/99

% Response

0-10 10-20 20-30 30-40 40-50 50-60 60-70 70-80 80-90 90-100

% of total solar system mass

Which Last Longer: Low- or Hi-Mass Stars
Asked on 12/16/99

% Responses

Low Mass   Mass is Irrelevant
      High Mass   I Don't Know

Earth w/ .5M$_\oplus$:Orbit Period Change?
Asked on 10/26/99

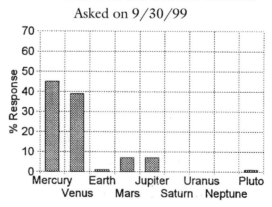

Planet Physically Most Like Earth
Asked on 9/14/99

Which Planet Has the Hottest Surface?
Asked on 9/30/99

FIGURE 6.2 Results of misconception-based attendance questions—Earth and other planets.

Earth w/ .5M$_\oplus$:Orbit Period Change?
Asked on 12/16/99

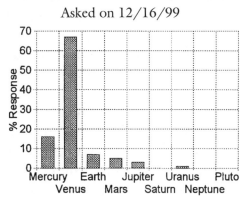

Planet Physically Most Like Earth
Asked on 12/16/99

Which Planet Has the Hottest Surface?
Asked on 12/16/99

You may notice that most of the answers to the questions shown here have an intuitive, common-sense, or common-knowledge, albeit incorrect, answer:

> The cause of the seasons is the changing distance between the Earth and the Sun. (As we have seen, it's actually the tilt of the Earth's rotation axis.)
>
> The lifetimes of higher-mass stars are longer than the lifetimes of lower-mass stars. (As we have seen, lower-mass stars actually last longer.)
>
> The orbital period (year) of a lower-mass Earth would be shorter than the year today. (As we have seen, a planet's mass actually does not affect the period of its orbit around the Sun, which is the length of the year.)
>
> The planet most like the Earth is Mars. (Actually, Venus is more similar in size and composition, although Mars is more similar in rotation period.)
>
> The planet with the hottest surface is Mercury. (As we have seen, Venus is actually hotter because of the greenhouse effect in its atmosphere.)

The question about the percentage of the solar system's mass contained in the Sun is interesting because most people state an initial belief, but there is no *common* one. Initial beliefs range from much less than one percent up to over 99 percent. (The actual value is a tad over 99.85 percent.)

The fact that the answers to the question about the orbit of a different-mass Earth were particularly resistant to change provides a clue about what adopting some new beliefs can require, namely, understanding the underlying mathematics. I teach this astronomy course with very little math. Manipulating equations to prove that the mass of a body has no bearing on how long it takes to complete one orbit around the Sun would be beyond many of the students. Even students in more advanced astronomy classes *who have not yet seen the mathematics* get this question wrong. However, after they work through the math, most of them get it right.

The number of students who held incorrect prior beliefs about some of the questions is actually slightly larger than the number I collected. I discovered this after hearing two of the groups that stayed after class discuss their answers. Students in each group said that whenever the question had only two possible answers (as in: "Does the Moon rotate?" "Which last longer—high-mass stars or low-mass stars?" "How does the Sun compare in size to other stars?"), they wrote down the one they didn't believe *and got it right*. This is part of the "street smarts" that some students develop, of course. Keep in mind that I tell them repeatedly that the answer will only be used to help me understand their prior beliefs; it will not be held against them and will never even be compared to their names. That some students still try to get "the right answer" suggests that they avoid addressing their incorrect beliefs by treating the questions as a test rather than examining what they thought. This is only one of a variety of defense mechanisms people develop instead of reevaluating what they believe. Other responses that protect core beliefs include ignoring new information, consciously or unconsciously rejecting it, interpreting it in a way that fits prior beliefs, allowing it to make only peripheral changes, and filing it away for future reference.

That students had a variety of emotional responses to these questions during the semester made me curious to know what they thought about my asking them in the first place. Therefore, my last question of the semester was a request for their comments on whether the questions had affected their learning experience and thinking process (question 22, above). Here are some typical answers.

> "I think they have been *really* [student emphasis] cool. It's funny the things that you think are right but aren't."
> "Yes, they have showed me how little I know."
> "Yes, for the most part I did not know the answers. From these I learned something I will remember."
> "Yes, they make me think on my own and apply what I have learned throughout the course."
> "Yes, I always remember the answers after—they're stuck in my mind (especially if I get them wrong)."

"Yes, I discovered how naive some of my thought patterns have been. It's almost funny. I've learned *a lot* [student emphasis] in this class. Thanks."

"These questions have been useful because the suspense makes them memorable."

"Yes, they made me question some of the things I believed were true."

"Your questions have brought answers to many questions of my own."

"Yeah, because they tell me what [sic] little I know!"

"Yes, one learns from others' mistakes."

"Misconceptions have been helpfull [sic] to me."

"I found these questions to be fun. I enjoyed hearing the answers [in the next class] after I tried to come up with my own."

"I think they have. It's always a nice thing to talk about with a friend or a relative. Continue doing it!"

"I think they have. They tend to stick in my mind after class and in general I tend to remember them for a long time."

"Yes. I believe so. It has kept me interested in the course material more than I probably would have been."

"I think these questions have been very thought provoking."

"Good because after we get all that information [in lecture], we get a question to chew on."

"Yes, these questions have been very interesting. Keep it up!"

I have also received some neutral responses, like "they didn't help," but I have yet to receive a negative comment. As you can see, there are many dimensions to even this one aspect of the educational experience. To grossly understate the point, the attendance questions caused more students to spend more time talking together about the course than they would normally. In many ways, the process transformed what had been a normal large lecture experience into a more dynamic and instructive experience. Small groups crystallized. Students realized that they were not alone in their incorrect beliefs. Indeed, this large group experience showed them that since virtually everyone had numerous incorrect ideas about the cosmos, it was "okay" to acknowledge that they harbored incorrect ideas. Some students were comfortable arguing for their beliefs with each other, with my answers at the beginning of the next lecture serving to arbitrate between them.

By discussing the questions, and later the answers, students take steps to evaluate some of their deep-seated beliefs about the natural world. They also transform the traditional large lecture experience into something else, a learning experience driven in part by their own beliefs and their own desire to understand.

Neither asking misconception-based questions nor any other single activity will correct all the disadvantages of large lecture classes. However, this example shows that a much more effective learning environment can be created, even within the restrictions of large numbers of students and small numbers of teachers. By the way, the idea that asking the question would keep class attendance high proved valid. Attendance remained very high throughout the semester, rather than falling off dramatically after the first test, as it had done historically. Furthermore, the weaker students did significantly better than usual.

Filling the Void

Once you accept that one of your deep-seated beliefs is incorrect, you immediately begin to formulate a new understanding of that concept. Without sufficient accurate information or a better understanding of the subject than you had when constructing your original idea, you are quite likely to create another incorrect or incomplete belief. Indeed, you might well re-convince yourself of your original belief. This is common—you hear something new that corrects your ideas, but without hard confirmation and a compelling alternative, you revert to your old way of thinking.

A very short time, perhaps a few days, exists between letting go of old ideas and acquiring new ones or reacquiring old ones. This small window is also the best time to learn correct science, either on your own or with the guidance of a teacher. Suppose, for example, you believe that Polaris is the brightest star visible in the northern sky and I tell you the brightest visible star is actually Sirius. If you don't see proof of this, chances are you will soon forget about Sirius. You are much more likely to remember this fact if you actively investigate it. I invite you to go out the next clear night, find Sirius, and compare its brightness to that of Polaris[1] and the other stars in the sky. You can easily find Sirius whenever Orion is "up" at night. Locate Orion's belt and, extending it, follow a straight line down (southward) and to the left. The first bright star you will encounter on that line is Sirius. You can find Polaris by first locating the Big Dipper, finding the two stars farthest from the handle, and then following a line upward (out the top of the pot) from them to the first moderately bright star. You can easily see that Sirius is brighter than Polaris by comparing them. Indeed, Sirius is brighter than all the other stars in the sky. Going through this *process* greatly reinforces the correct belief, in part because you see the facts for yourself and in part because you are actively using different parts of your brain to achieve this realization.

This type of hands-on discovery process is extremely effective in making people face the invalidity of their beliefs. It is analogous to the experience you may have had in a spelling bee or on a show when you must commit to an explicit answer. In that setting, you are acutely aware of what you think, and when you're wrong, the public correction you endure will motivate you to remember the correct answer.

Often it isn't possible to do a hands-on experiment or make an observation, such as in a large lecture course. The next best thing is to see presentations of experiments or other concrete evidence that directly contradicts your beliefs. For example, if you believe that all moons are spherical, like ours, this belief will be called into question the first time you see a picture of Mars's moons Phobos and Deimos (figure 6.3).

---

[1] A northern hemisphere observation.

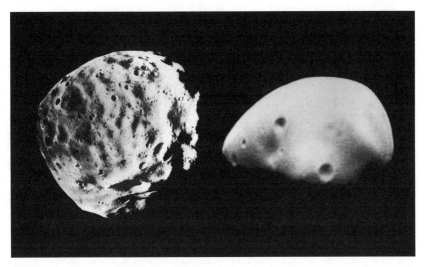

FIGURE 6.3 Moons of Mars. Phobos *(right)* and Deimos. Neither is remotely spherical.

BLIND SIDING

Knowing common incorrect beliefs and how they affect your reasoning are the first steps in confronting and replacing them. Getting past our instinct to defend what we believe to be correct is another important issue. A teacher might say something like, "Many people believe that the only function a telescope has is to magnify objects. Let's see what it really does." You can probably already see the problem with such an approach: students immediately get defensive and go to extraordinary lengths to protect their threatened belief. As just noted, a more successful approach is to guide the students to discover inconsistencies between their beliefs and how nature really works. Another, less direct but less equipment-intensive, way is to get behind the defenses: teach directly to the misconception.

Suppose, for example, that you were learning about Saturn's rings in my class. For many people, the word "ring" creates the belief in a solid ribbon orbiting that planet. Images taken through small Earth-bound telescopes certainly support the belief that the rings are solid, perhaps with one or two gaps (Cassini's and Encke's divisions). It is true that high-resolution photographs of Saturn's rings taken by the Voyager spacecraft show numerous concentric "ringlets" of different brightness, but as yet we don't have any photographs that show individual objects making up the rings. So it is not surprising that many people believe Saturn has a few rings that are solid all the way around the planet.

By this time in the course, you would already have learned Kepler's laws (orbits are elliptical; the farther a planet is from the Sun, the slower it is moving in its orbit and vice versa; and the period of a planet's orbit is directly related to its average distance from the Sun). Saturn's rings obey Kepler's laws because they orbit under a gravitational force, just like the planets orbiting the Sun. Observations reveal that Saturn's main ring system is a staggering 30,000 miles wide, over three times the diameter of the Earth. Using Kepler's laws, you can calculate that the inner edge of the ring system is rotating nearly twice as fast as the outer edge.

Suppose now that Saturn's rings really were solid. The fact that the different parts of such wide rings would be moving at such greatly different speeds indicates that they would quickly shear apart as the inner part of each ring ripped ahead of the corresponding outer part.

"But what if the ring is solid and strong enough to hold itself together, either by its own gravity or because it is made of a strong material, like iron?"

Good points, well worth considering. Observations made of the rings edge-on reveal that they are no more than a few kilometers or miles thick. Since they are that thin, they lack sufficient gravitational force to stay together as a single ring, especially under the shearing force created by their orbital motion, as spelled out in Kepler's laws. Furthermore, if a solid ring were held together in the same way that a single rock or sheet of iron on Earth stays together (by electrostatic forces between the atoms), the ring would also start vibrating, like the head of a drum, due to the changing gravitational tugs of Saturn's moons. The vibrations would cause such a thin ribbon to tear apart in a matter of days. Furthermore, debris falls through the rings toward

Saturn. This material would have caused a solid ring to age, crack, and disintegrate long ago.

Seeing what processes are actually taking place, rather than just hearing that the rings can't be solid (in this example), presents a more compelling case for believing that the rings are made of many smaller particles (typically pebble- to boulder-sized). Now you have the physical information necessary to help you construct a new, more accurate belief. If argued well, the position that planetary rings are not solid is also strengthened by the teacher or other source of information undermining many[2] of the relevant ideas that you use to hold on to the incorrect belief.

To summarize this way of teaching, it is important to know the answers to the frequently asked questions (FAQs) about topics that carry common misconceptions and, ideally, to answer the questions before they are raised. By doing this, teachers provide their students with new, correct constructs that will be readily available when the students begin examining their prior beliefs (such as solid rings) to help disassemble these misconceptions and construct new, correct beliefs.

INCONCEIVABLE

The effect of truly changing a deep-seated belief is quite interesting. It often becomes virtually impossible to imagine that you once held an alternative belief. I once read a paper from which I learned that some children have a neat way of reconciling the difference between their everyday perception that the Earth is flat with the "fact" they are told by their teachers and other trusted sources that the Earth is round. Instead of throwing out their flat Earth model, they combine it with

---

[2] For most concepts, most people hold one of several incorrect beliefs and relatively few people hold one of a few outlying misconceptions. In such cases, knowing the common incorrect beliefs and their foundations makes this approach possible. However, some subjects are sufficiently far from everyday experience that justifying them leads different people to numerous different misconceptions that are much harder to address completely. For example, start asking people what causes an aurora and you will get *dozens* of (incorrect) explanations.

the round Earth concept to create a disk-shaped Earth. It's flat where we live and has a round edge.

Intrigued, that evening I asked my son Josh to describe the shape of the Earth. On a piece of paper he drew a rough circle and a stick figure in the middle. The next day I asked my astronomy class if they had ever thought of the Earth as a flat disk. A few people raised their hands, no more than half a dozen out of 250 students. However, over the next week, several students drifted into my office with similar stories. They told their parents about this flat/round Earth concept, and their parents replied that as children these same students had believed in this or a similar version of the Earth. The students all had forgotten they had ever believed it.

Why do people often forget previous beliefs? When we have to replace deep-seated beliefs that have been irrefutably shown to be wrong, our minds go through dramatic readjustments. These include reorganizing information in ways inconsistent with our prior beliefs. Our new reasoning takes us directly to our new beliefs, making it hard to reconstruct the old ones.

Tabula Rasa

Let me end this chapter on unlearning misconceptions with the story of a student I'll call M.F., who took introductory astronomy from me. Early in the semester she came into my office and explained that she had never been interested in astronomy. Even as a child, she had not cared about the night sky. She was a journalism major and was taking astronomy only to fulfill her science requirement. "Nothing personal," she said, "but I don't know how you can stand to think so formally all the time." She wanted to know how much math she would need and whether there were tutors available, as she did not feel comfortable with her ability to think about "cut-and-dried" scientific things. I assured her that the course was entirely descriptive and that the university did have tutors available. I was more amused by her misconception that science is "cut and dried" than by the fact that she had never been interested in astronomy.

M.F. came to see me with questions before each exam. They were typically about how much detailed information she needed to know. She was taking the course very seriously, organizing her notes into patterns that were particularly useful for her way of learning. The second clue (after the fact that she had never been interested in astronomy) that something about M.F. was unusual came as she turned in the final exam. Recall that I give the students the chance to get extra credit by turning in 42 incorrect beliefs that I have corrected for them throughout the semester. Handing me her list, she told me that she wanted to get the extra credit but could only think of a dozen incorrect beliefs. I'd never encountered a student who had any difficulty identifying 42, much less 12. The kicker came when I sat down to determine the final grades. I began with the raw exam scores, which I curved as necessary. Then I added in the extra-credit points. Finally I printed out a spreadsheet showing all the grades, so I could go over each student's scores individually, looking for marginal cases that needed special consideration. The column of computer-calculated course grades had mostly double-digit scores, with occasional numbers ranging from 100 to 104. Numbers over 100 were consistent with the extra credit and the increased grades on tests with curves. Out of curiosity, I looked to see who had gotten the single 104. It was M.F.

Surprised, I scanned her individual grades. Out of 250 test questions throughout the semester, she had only gotten 2 wrong. How could this be? I must admit the first thought I had was that she had cheated, perhaps by sending a ringer in to take her exams. But instantly I rejected that because I recalled her handing me at least two exams, including the final, on which she got a perfect score. Then the logical explanation suggested itself: she came into the class with few misconceptions or other incorrect beliefs about astronomy. Therefore, the information I gave didn't have to compete against incorrect knowledge already residing in her brain. She was able to create a correct image of the cosmos essentially from "whole cloth" and use this knowledge to answer both rote-memory questions and questions that required some reasoning. She vividly showed that without misconceptions vying for your attention, it is much easier to acquire a sound scientific foundation for understanding nature.

7

Let the Buyer Beware

HOW TO AVOID
FUTURE MISCONCEPTIONS

Easier Said Than Done

We are bombarded with information every day, some of it clearly valid, some of it clearly false, much of it suspect. Just as there are "knock offs" (cheaper, lower-quality copies of high-quality products) in the marketplace, you can't take much of what you hear or see these days at face value. When we accept inaccurate information as correct, it frequently leads us to develop erroneous beliefs, which in turn can cause us to draw invalid or incorrect conclusions and to make bad decisions. Most advertising falls into this gray area. Consider a kind of claim you often hear in advertising: "Studies show that our product is the most effective pain reliever you can buy." This statement is completely misleading. A variety of techniques exist for dealing with new information; before considering them, let's see why such advertisements should not be trusted.

The primary problem with such claims stems from the glib use of the word "studies."[1] For all the advertisers tell us, these studies could have been done by asking different groups of employees at the pharmaceutical corporation that manufactures the drug which pain reliever they find most effective . . . not exactly an unbiased sample. To be meaningful, studies and experiments must be scientific, double blind, placebo controlled, and statistically significant, and many of them must be run by independent researchers.

A *scientific* study or experiment must take into account as many factors that could affect the results as possible. In the case of pain-reliever tests, researchers must consider things like dosage, patient history, other medications the patients are taking, and the various causes of pain. In a *double-blind* study or experiment, neither the people participating in the study nor the testers administering it know which medication or placebo each person is taking. This same technique also applies to side-by-side comparisons, such as taste tests. Using double-blind experiments helps eliminate irrational biases that either users or testers might have toward one product or another. *Statistically significant* studies involve testing enough people using the product, alternatives, and placebos to come to meaningful conclusions. Furthermore, it is essential that *independent researchers* test ideas or products. As I mentioned earlier, science is a verifiable, repeatable process, and studies by many different groups are necessary to assure the validity of any scientific assertion. Independent and repeated testing helps minimize the possibility that experimental design and experimenter biases, not to mention outright fraud, will affect the outcome of the experiment. With the ad just discussed, clearly there is no way for the consumer to know if the cited studies satisfy all these criteria or if enough of them were done to support the assertion.

---

[1] Likewise, the word "pain" is suspect. We suffer different types of pain from different sources, often requiring completely different types of medications to control them.

The issues raised in this example are typical of those we should all consider whenever we hear or see "scientific" claims. The insights we can get from asking such questions reveal the value of critical (or skeptical or scientific) thinking and help us avoid acquiring new incorrect beliefs. As with so many terms, "critical thinking" has many definitions. I use it to mean the evaluation of the value, accuracy, or authenticity of an assertion and the reasoning that leads to a supportable decision, belief, or action.

Unfortunately, critical or skeptical thinking has received a bum rap in many circles, because these words often generate negative feelings. When we think of someone as critical, we often mean that they go out of their way to find fault with things. When we think of someone as skeptical, we mean that they doubt everything they hear. While critical thinking or skeptical thinking sounds confrontational, in reality such activity is truly constructive because it helps us create valid beliefs and prevents us from adopting invalid ones.[2]

Here, then, are some guidelines to help you think critically, decide about the validity of new claims you hear, and thus avoid developing new incorrect beliefs. I discussed some of these earlier, but I think it is useful to have them all in one place.

DO IT YOURSELF

1. Do not rely on unrepresentative, biased, or incomplete data to prove anything.

If the only people interviewed about their favorite pain reliever were employees at the pharmaceutical company, they make an unrepresentative and biased sample. They are unrepresentative because they don't

---

[2] My three favorite books on the subject of critical thinking are: *How to Think About Weird Things: Critical Thinking for a New Age*, second edition, by Theodore Schick Jr. and Lewis Vaughan; *The Demon-Haunted World: Science as a Candle in the Dark*, by Carl Sagan; and *How We Know What Isn't So: The Fallibility of Human Reasoning in Everyday Life*, by Thomas Gilovich. These books, among others, provide even more insights into remaining centered in an unbalanced world.

comprise a meaningful cross-section of the public. They are biased because they have a vested interest in the success of the drug—their jobs may well depend on it.

Incomplete data means not enough results from enough experiments, studies, or subjects to draw a valid conclusion. Related to this is the problem that individual "case studies," "medical studies," and "personal experience" also provide insufficient evidence. Several times my family doctor has told me that he believes some remedy is effective just because he has seen the symptoms of several of his patients improve after they have taken it. He is unjustified in drawing such a conclusion based on so few cases, studied unscientifically.

You will often hear about or read "case studies" in which a patient was cured while taking a new drug or a group of children did better than average on a standardized test using a new teaching system. Such single cases are not sufficient to justify the conclusion that the drug or teaching system will be successful in general, or even the conclusion that they were successful in these specific cases. One problem is that people who know they are being studied or evaluated often do better than average, regardless of the technique used on them. An early example of this effect occurred at the Hawthorne manufacturing plant run by Western Electric in Cicero, Illinois. Between 1927 and 1932, the workers at this plant were told that they were involved in a study of production efficiency. Then, whatever the experimenters did, such as change the lighting, humidity, or break schedule, productivity rose.[3] Similarly, in hard sciences like physics and astronomy, the results of single observations or experiments, conducted with the intent to test the validity of a particular theory, are not sufficient to convince other scientists that a certain phenomenon exists. Every observation and experiment must be replicated by independent researchers and the reasons for the results must be understood.

We all rely on our personal experiences to draw conclusions, but doing so about things scientific is often a mistake. If some remedy

---

[3] The phenomenon of change due solely to the participants' knowledge that they are being studied, rather than to something substantial, is now called the Hawthorne effect.

seems to work for you, chances are you are going to tell your friends about it and advocate it if the need arises. The problem is that what seems to work for you or me may not actually be the curative agent, and even if it is, it may not work for most other people. You need to have enough scientifically tested information in order to make valid claims.

2. Question authority.

Just because someone in authority says something doesn't make it so, because authority figures are often not experts on the issues about which they speak. Grade-school teachers, news broadcasters, and politicians come to mind. If a politician rails against spending on something scientific, alleging that experts have assured him or her that the project is not worth the money, you can be pretty sure that the project wouldn't bring any money into the politician's district or state. In other words, under some circumstances claims are made about scientific matters without regard to their merits, importance, reality, or promise of further insights.

3. Consensus of expert opinion (in the experts' field) should be used or considered correct until shown to be otherwise.

If issues are important, it is worth learning about them directly or getting expert opinions on them. Experts should be treated as reliable sources of knowledge rather than as authority figures. Unlike authority figures, who implicitly expect you to believe what they tell you without further questioning, experts aren't trying to force you to believe anything. Experts share their knowledge primarily to keep you informed and perhaps to gain some publicity.

Scientific experts are people who have had training and experience in some field and who have gone on to do research and publish papers that have successfully undergone peer review, a process by which the results of scientific work are examined by independent experts who judge them on their scientific merits. This is why "experts" are the people best qualified in the world to discuss their fields.

A problem that often occurs is that experts disagree about results in their field. Differing scientific interpretations of experiments, observations, and theories often have their own expert advocates. That is why, just as it is necessary to do many scientific studies to validate claims, it is often necessary to wait until there is a majority expert consensus about a scientific matter before you develop a belief about it. (This is, of course, easier said than done, since when we want to know something, we want to know it now.)

For example, in searching for purportedly massless particles called neutrinos that are created inside the Sun, experiments beginning in the 1960s detected less than half the number of neutrinos predicted by the theory of neutrino production rates. This led some experts on solar physics to claim that there was an error in our understanding of the theory of nuclear fusion (which predicts neutrinos and how copiously they are created inside the Sun). Other experts claimed that the theory of fusion was correct, but that the properties of neutrinos were misunderstood and consequently the techniques used to detect them needed refinement. Both of these explanations had far-reaching consequences for our understanding of nature, so further experiments were undertaken. Decades later, these improved experiments revealed that our understanding of fusion was just fine, but neutrinos did have different properties than originally believed. As a result of this recent work, there is a growing consensus that neutrinos have mass, which, in turn, has had a major impact on our understanding of the evolution of the expanding universe.

Even when consensus is eventually reached on scientific issues with controversial evidence or evidence open to different interpretations, a minority in that field will still hold alternative views. For example, while a consensus is now forming that the universe will expand forever, some astronomers and astrophysicists still believe that it will someday recollapse. Some of the latter believe, plausibly, that there isn't enough evidence to support the view of permanent expansion, while others hold that the universe should eventually recollapse for religious or aesthetic reasons.

Minority scientific viewpoints are often the starting points for nonscientific beliefs. For example, in 1989 two University of Utah researchers claimed that they were able to cause atomic nuclei to combine (fuse) under conditions that the mainstream scientific community believed were unsuitable for such events. This concept became known as cold fusion. Subsequent attempts by independent researchers to repeat the experiments all failed, although interesting new chemical properties of matter were observed that could explain the initial experimental results. To this day, some people still believe that cold fusion is possible. This leads them to advertise cold fusion-related books, magazines, equipment, and memorabilia. Nevertheless, cold fusion has never created any energy, and I believe it never will.

Keep in mind that experts in one field are not necessarily experts in other fields, and therefore their beliefs in those areas should be taken with care. For example, when a reputable cosmologist persuades a country to invest millions of dollars in geological programs that the vast majority of professional geologists find dubious, one must ask whether the person's reputation or the science in the project was the persuading factor. Along the same lines, you have every right to know the credentials of "experts" to help you decide just how much to believe them.

4. Don't rely on the words of the ancients.

This is a corollary of "question authority," but it is worth discussing separately. Just because a theory dates back centuries or millennia doesn't mean it is correct. Furthermore, there is no scientific evidence that ancient philosophers or theologians had access to information about natural laws that isn't available to us today—if we search for it. So if an ancient authority figure (say, Aristotle) asserted something (that more massive objects intrinsically fall faster than less massive ones), we must reject the historical assertion if experimental evidence proves otherwise (as first shown by Galileo in the seventeenth century, objects with different masses fall at the same rate). Indeed, I have often wondered why Aristotle didn't do a simple experiment or two and convince himself that his idea was wrong. Perhaps he did, but under the power of his own misconception, rejected his experimental results. Or perhaps air friction confounded his results.

5. Traditions and traditional beliefs are insufficient justification for believing that something is true.

There is a tradition in my family going back four generations that we are descended from a famous theologian, Zvi Hirsch Kalischer. I learned of this belief in the early 1990s when I became interested in my genealogy. Some of my newly discovered cousins and I tried very hard to prove or disprove this story. But Kalischer lived in the early part of the nineteenth century, and a variety of historical and social traditions and barriers between then and now have prevented us from getting to the truth. Even Kalischer scholars have been unable to give us a definitive answer.

The question is, then: Should we believe this comfortable story without proof or should we not? My feeling is that we should not. In the first place, if we propagate this belief based only on hearsay, future generations will have even more conflicting and unreliable evidence to cull through when they try to determine the truth. In the second place, it feels presumptuous to tell certified Kalischer descendants that we are related to them when we may not be.

Science, however, can provide a high probability answer to the question of my family's heritage through DNA analysis. By comparing my DNA to that of a known Kalischer descendant, we can find out whether this family legend is true. The point is that many traditional beliefs are now open to scientific verification. Whenever possible, use such methods to help establish a firm basis for belief.

6. Consider the arguments rather than the person providing the information.

Because you can get solid information from distasteful or disreputable sources, judge all data on their own merits. During the Cold War, some Soviet scientists were often scorned in the West, in part because they were helping "the enemy" develop weapons of mass destruction. Nevertheless, what these people were doing was real science. Therefore, unless Western scientists wanted to take years to reinvent things, they studied the work of their Soviet counterparts strictly on its own merits.

As another example, though his scientific work is respected and his ideas are widely accepted, Sir Isaac Newton was probably one of the nastiest people of his time. If someone got on his bad side, Newton went out of his way to discredit the person, or worse. Nevertheless, Newton's contributions to science stand as towering intellectual achievements.

I grant you that it is often very hard to separate the message from the messenger. Listening to people I find distasteful talking about things of substance, I often ask, "What is this person's real agenda?" Trying to fathom this takes a lot of intellectual energy away from the scientific, engineering, or other type of concept the person is presenting.

7. Use Occam's razor whenever you have two or more competing concepts explaining the same thing.

Just to restate the principle of Occam's razor: If you learn two or more reasons why something occurs, choose to believe the one that requires the fewest unproven assumptions. Suppose, for example, you learned that pulsed radio signals had been detected from a wide range of locations in our galaxy and that the spacing between the pulses from each source was more regular than the ticking of the best clocks here on Earth. At first glance, the precision of the pulses coming from a universe filled with otherwise random radio signals certainly invites consideration of intelligent intervention. Indeed, this was the first thought that went through radio astronomers' minds when such objects were first discovered in 1967 by Jocelyn Bell Burnell in Cambridge, England. One possible explanation for the pulses is that they are radio navigation beacons constructed by aliens to expedite their travels around the Milky Way. Here are some issues surrounding alien construction of such signals. First, the amount of energy emitted by them is "astronomical," requiring very powerful sources. This is an engineering, rather than a scientific, issue. Next, the technology for creating such precise pulses is challenging to create. Again, an engineering issue. Third, the implication of the distribution of these sources is that if aliens created them, then the galaxy is pretty fully inhabited by other life forms.

Now consider a second explanation—that the signal sources are very compact, rotating remnants of stellar evolution, objects called neutron stars, containing magnetic fields that do not pass through the star's rotation axis. As those magnetic fields whip around the axis, their changing directions stimulate the emission of radio waves in nearby interstellar gas. Astronomers have developed sophisticated models describing the formation of neutron stars from stars initially containing between eight and sixteen times the mass of the Sun. Using the Sun as an example of a rotating star with tilted magnetic fields, we believe that many neutron stars have similar properties. Calculations show that neutron stars have more mass than the Sun but are compressed into volumes of space only a few tens of miles across. Such objects are physically capable of rotating fast enough to give off the observed pulses with the observed precision.

Occam's razor dictates that we accept the second explanation for these so-called pulsars (discussed in chapter 2). By doing so, we don't have to allow for the existence of as-yet-unidentified intelligent life forms, the "unproven assumptions" in this example.

8. Question assumptions.

Whenever you try to understand anything, you necessarily make some assumptions. If you use the wrong ones, you will come to invalid conclusions and therefore create misconceptions. For example, consider this argument: "Life in the solar system first formed in water. That water must have been liquid to allow for enough molecular interaction in it. For the water to be that warm, it must have gotten enough energy from the Sun. Therefore, life must form only in the inner solar system where the Sun provides sufficient heat—no farther out than Mars."

In this argument, I have made several assumptions about the origins of life in the solar system. I began by assuming that the life that exists here originated here, rather than having developed elsewhere in the Milky Way galaxy and then drifting over and blossoming. While this is likely to be correct, it has yet to be proven and is open for debate. Next, I assumed that life forms in water. Is that true? Again, we don't know yet, but it is the majority view among evolutionary biologists. Then I asserted that the water in which life first formed must be liquid. Is that true? At this point, it is indeed the majority view among evolutionary biologists. Fourth, I assumed that liquid water in the solar system gets its energy to be liquid from the Sun. That seems plausible, and it is true for the water on Earth. But in fact it isn't always true.

There is very strong evidence that Jupiter's moon Europa has liquid water under its solid surface. But at that distance the Sun is too far away to supply enough heat to melt water, much less to keep it liquid for billions of years. The temperature on the surface of Europa is -100°C (-150°F). The heat necessary to keep water liquid there actually comes from the changing gravitational pull of Jupiter and its other moons, Io and Ganymede. When Io (which is closer to Jupiter than Europa) is on the same side of Jupiter as Europa, it pulls Europa toward it (and Jupiter). When Ganymede (which is farther from Jupiter than Europa) is on the same side of Jupiter as Europa, it pulls Europa farther from the planet than usual. These changes in distance from Jupiter actually cause Europa to change shape, for the same reason that the Earth's oceans move in response to our Moon's location. This so-called tidal effect creates friction inside Europa as its atoms move relative to each other. This friction, in turn, generates heat that keeps Europa's internal water liquid.

This point about valid assumptions is one of the key rules of reasoning. If you use an incorrect assumption, your conclusion will not be valid.

9. Be sure that the logic you use is correct.

Logic is the set of rules for valid reasoning. If you use correct logic and valid assumptions, then you will come to correct conclusions. Perhaps the simplest rule of logic is the transitive one: Napoleon was French. All French are European. Therefore, Napoleon was European.

Consider the following extreme example of drawing an invalid conclusion from a correlation between two things: People who are successful drive expensive cars. Therefore, the best way to become successful is to buy an expensive car. This type of incorrect logic has the technical name *post hoc ergo propter hoc*, which means "whatever happens after this must happen because of it." In everyday terms, it is a case of putting the cart before the horse. Logic has an extensive and

well-defined set of rules and guidelines, many of which are all too easy
to inadvertently violate.

*10. Just because something is logically possible doesn't mean it
necessarily occurs.*

Logical impossibility prevents some things from occurring. For example, the rule of noncontradiction says that something cannot have a
property and lack the property at the same time—a person can't be
both living and dead. On the other hand, many things are logically possible but physically impossible. For example, it is logically possible for
someone to run a mile in one minute. However, to do this means running at sixty miles an hour for a minute, which is physically impossible.

*11. Just because something is physically possible doesn't mean it
necessarily occurs.*

Carrying the previous point one step further, things that can happen
don't have to happen. It is both logically and physically possible for a
collection of atoms to spontaneously come together and form a rose.
However, calculations of the interactions involved strongly indicate
that such an event is extremely unlikely ever to occur.

*12. Just because something hasn't been disproven doesn't mean that it is
correct or valid; conversely, just because something hasn't been proven
doesn't mean that it is false or invalid.*

The efficacy of countless over-the-counter remedies available today has
not yet been scientifically proven. Testimonials, word-of-mouth support, anecdotal stories, and hearsay keep them going. However, no
such claim for these remedies is valid evidence that any of them works.
After undergoing scientific studies, some of them will eventually be
proven effective and some will be proven useless or even dangerous.
Until then, people will continue to buy them but should not expect
the support of the established medical community.

Back in 1998, a product called "Vitamin O," which apparently contained just salt water, was advertised as a treatment or preventative for

cancer, heart disease, and lung disease. The distributors made false claims that the value of the product had been established by medical and scientific research, and they used alleged testimonials about its effectiveness. For over a year they allegedly sold more than 60,000 vials of Vitamin O at $20 or more per vial each month. A physicist, Robert Park, exposed this product as useless, and within less than a year the U.S. Federal Trade Commission fined the company and forced it to cease operation. Nevertheless, by the time production of and advertising for Vitamin O were officially stopped, it had brought in over $10 million.

13. Be extremely wary of common sense and intuition when applied to scientific matters.

Just five hundred years ago, common sense told virtually everyone that the Earth was the center of the universe. Just two hundred years ago, common sense told everyone on Earth that it was impossible to instantaneously talk to someone on the other side of the world, or even to someone just a few miles away. Today, virtually all of us take this (via telephone, of course) for granted. Beware of believing arguments simply because they seem "plausible." Common-sense ideas must be tempered by reference to scientific realities, which often seem counterintuitive at first.

Most people have a fair amount of faith in their intuition: "She seems like a nice person," "I trust that doctor." In reality, of course, some of us have better intuition than others. As I tried to show earlier in the book, very few of us have good intuition about the workings of nature, and we should not trust our "gut feelings" in that realm. Even scientific experts believe their intuition only so far; we perform calculations or make experiments or observations to test our ideas and beliefs.

14. Don't find patterns where they don't exist, don't overgeneralize, and be wary of analogies.

Our ability to recognize patterns has proven very beneficial in our evolutionary history. Those ancestors with the ability to identify predators fastest were the most likely to survive. Hence, this ability has been hardwired into our brains. We quickly classify things that look remotely like human faces as human until we learn otherwise. This gives us more time to determine whether they are friends or foes. Branches waving in the wind can look like a person threatening you as you walk by woods at night. Children see many threatening objects, like monsters in the shadows. The problem is that when we find patterns where they don't exist, we use those incorrect beliefs to come to incorrect conclusions.

In Bucksport, Maine on the Penobscot River is the iron-fenced Buck family cemetery. On the tombstone of Bucksport founder Colonel Jonathan Buck, a discoloration that *looks like* a foot and leg seems to be etched. This feature's apparent connection with something recognizable led to many colorful stories about its origin, including that when Colonel Buck, who was a local judge, sentenced a woman to be hanged or burned at the stake as a witch, she swore a curse against him. After Buck's death, the shape of a foot and leg appeared, allegedly as part of the curse. (There is actually no record of anyone being hanged or burned for witchcraft in Maine.) I must admit that I am dubious about these explanations. I prefer to think that the mark is a natural flaw in the stone.

In science, many things look similar but in fact are completely unrelated. When learning something new about the natural world, be wary of the absolutely normal responses, "That reminds me of . . ." or "That's just like . . .." Finding connections or making comparisons that don't really exist can lead you to unjustified conclusions. The origins of fire on Earth and the heat and light from hot gases of the Sun are completely different. Unfortunately, astronomers confuse this issue for nonastronomers by using the expression "burning fuel" to describe the fusion in the Sun that creates its light and other energies. Similarly, though they are called the same thing, the tails of comets and meteors have different origins (and different directions relative to the comet and meteor bodies); and as we have seen, comet tails rarely trail comets.

15. Identify and investigate the core issues.

On a public radio show called Car Talk, two brothers field calls about car problems. On more than one occasion someone has called up and begun by stating their car's make, year, mileage, and color. The first three pieces of information help the hosts, Tom and Ray, figure out what is wrong. However, the color of the car is completely irrelevant to virtually all the problems that might beset it. When we learn something new about the natural world, some aspects are essential to our understanding, while others (like car color) make individual cases distinctive but do not provide insights into the underlying concepts.

Another crucial factor in understanding the natural world is separating scientific and engineering issues. As you may recall, when the Hubble space telescope was launched in 1990, astronomers discovered that its images were fuzzy. As a result, the amount of science accomplished during the first three years of Hubble's time in space was much less than had been hoped for. However, the reason for this shortcoming was entirely an engineering issue—an incorrectly shaped mirror. Needless to say, the science accomplished since the optics were fixed has more than made up for the years of ridicule NASA had to endure.

Engineering issues often come to dominate over science in decisions about whether to pursue some field of study. The question frequently boils down to cost. Many experiments and observations of the natural world now require complex and expensive equipment. Often it is hard to justify the expense because there is no guarantee that the knowledge gained from a multibillion-dollar investment in hardware will lead to any usable technology. In that sense, some big science projects are indeed gambles. Experience indicates that such gambles are often worth taking; worse come to worst, they only lead to engineering innovations that eventually contribute to our quality of life. And one thing is certain: if we don't make the effort and investment, we are *assured* of not finding any potentially useful new science that such experiments might reveal.

16. Develop intellectual humility.

Admit it when you don't know or understand things. Don't deceive yourself into believing you know what you don't or try to bluff your way through a situation that requires real knowledge. Ask questions. Consider this conversation I had with a geologist friend. Since I was speaking to someone with a similar amount of formal education, I assumed he had access to the same facts I did. Fortunately, he quickly corrected this assumption.

"As you know, there are an estimated 50 billion galaxies in the visible universe."

"Actually, I didn't know that. How do you come up with that number?"

"We have looked deep into the universe in a very small cone of space, back to the time when galaxies first formed. You can visualize this by taking an ice cream cone, cutting off the tip, and looking into the cone from the opening (the small end) you have made. We count all the galaxies in that cone of space and, assuming that there are roughly the same number of galaxies in every other direction, multiply the number of galaxies in that cone by the number of equivalent cones it would take to measure all the universe we're able to observe telescopically."

17. Be sure you have the correct definitions.

This is an example of intellectual humility that needs special attention. How often have you had a conversation in which people use either specialized or foreign words that you don't understand? Human nature often prevents us from asking what words mean—we're afraid of seeming ignorant. The cost of protecting our ego—bluffing our way through—is losing information and, often, keeping or developing incorrect beliefs. Suppose you hear someone state, "As you know, three planets have retrograde rotation," and you don't have a clue what "retrograde" means. Furthermore, you believe that "rotation" means something going around something else. Asking about retrograde rotation could gain you an explanation, like the following, and help you avoid acquiring new misconceptions.

Both "retrograde" and "rotation" have specialized definitions in astronomy. Many people think that all the planets spin on their axes (rotate) in the same general direction as the Earth. This spin direction is counterclockwise when seen from space above the Earth's north pole. Because the Earth's orbit (revolution) around the Sun is also counterclockwise as seen from the same vantage point in space, the rotation is called "prograde"—the rotation and revolution are in the same "sense." However, three planets don't have prograde rotation. Venus, Uranus, and Pluto all rotate in the opposite direction to their revolution around the Sun—they have retrograde rotation. Misunderstanding the meaning of words and then misapplying them can be as fatal an error as building a skyscraper on quicksand.

18. In science at least, the truth is not necessarily the opinion of the majority.

If most of the astronomers in the world believe that all the quasars in the universe are located more than ten billion light years from Earth, and if observations show that some quasars are to be found less than a billion light years away, then these astronomers are wrong in their belief about the distances to quasars. Opinions about the natural world must be based on scientific evidence, both theoretical and experimental, and even then, one must be open to new points of view based on better experiments, observations, and theories. The power of experiments and observations to show that experts are wrong in their beliefs is an accepted hazard of the trade. It should serve as a reminder to everyone that scientific beliefs are subject to change and that if you have come to believe something, even because of what reputable scientists have said (see point 3, above), you have to face the possibility that this belief is incorrect.

19. If you have good reasons to believe certain concepts and a new idea conflicts with those concepts, then you have reason to doubt the new idea.

Many of us are very responsive to new ideas. Word of a new diet can send thousands of people in a new culinary direction. New ideas on how to teach a subject can send thousands of teachers in a new educational direction. New ideas about how to travel to the planets can send thousands of engineers in a new rocketry research direction. New ideas about how the Sun's outer layers are heated up can send hundreds of astrophysicists (it's a small community) in a new research direction. However, it is always good practice to carefully compare a new idea with your existing beliefs before changing course.

Suppose you have been on a low-carbohydrate, high-protein diet and have lost a few pounds over the last year. Now you hear about a high-carbohydrate, low-protein diet that claims to help you lose weight faster. Should you reverse your previous belief system and try the new diet, or wait and see whether scientific studies give one diet the edge over the other?

Suppose you had been teaching mathematics to fourth graders using one method for several years, with adequate results on standardized tests. Then you hear about a "new math" that claims to give better results, as the New Math did back in the 1960s. It is healthy to be skeptical of these claims. You cannot imagine how many new teaching approaches are proposed each year and how many of them subsequently fail to produce the promised results.

Don't get me wrong. Sometimes you can take a gamble on something new and win big, as many people did in the "new economy" based on Internet companies. But even that bubble was fragile. More often than not, if you hear of something new that is at odds with something you have found to work, it is worth waiting until you have sufficient information to believe in the new thing before making a change. The same is true in science.

20. Keep an open mind when conflicting ideas exist, but not so open that your brain falls out.

I have just argued for a healthy skepticism about new ideas that are inconsistent with beliefs you hold. However, many people have beliefs about scientific concepts that seem plausible but really have no scientific evidence to support them. If you are confronted with a new belief and you really can't justify the one you have, then consider the alternative on its merits.

Most people know that the Earth's ozone layer is being depleted as a result of the use of stable chlorine- and bromine-based compounds, especially chlorofluorocarbons (CFCs). And most people reason (correctly) that the depletion occurs due to chemical reactions between the CFCs and the ozone atoms. What they can't figure out is how the ozone can be replenished once it is destroyed. Therefore, they often conclude that once the ozone is gone, it's gone forever, and we will eventually all be exposed to lethal doses of ultraviolet radiation from the Sun.

If you believe that the ozone layer will not be replenished, let me present an alternative belief for your consideration: it eventually will. In the short term, while the CFCs are still in the air, concern about increased surface ultraviolet exposure is very well founded—we *are* being exposed to more ultraviolet radiation than previous generations, which is why it is so important to use sunscreen. However, high-energy ultraviolet photons from the Sun not only penetrate the atmosphere to the Earth's surface, they also frequently bang into normal oxygen molecules (O_2) high in the atmosphere and rip the two oxygen atoms in those molecules apart. The freed oxygen atoms then combine with remaining oxygen molecules to create new ozone molecules (O_3). When enough of the CFCs and other harmful compounds are out of the atmosphere, the Sun's energy will rebuild the ozone levels to earlier, safer levels.

In this example, I have given an alternative idea about ozone replenishment and followed it up with a scientific explanation. When you hear alternative ideas to those you hold, demand scientific justifications and compare them to what you already believe. Which make the stronger case?

21. *Learn not to defend a belief farther than the evidence justifies.*

This is one of the hardest guidelines to follow because it goes to the core of our egos. It takes a lot of energy to defend something after you have run out of arguments to support it. Often defenses become irrational or even physical at this point. If you can listen to your arguments dispassionately as you make them, you can often get a sense of when you are repeating yourself, using verbal or physical force rather than the force of reason, or using invalid arguments to make your case. These methods are a sure sign that it is time to stop and listen.

Being able to listen to alternative points of view on their merits is often painful. Sometimes you don't agree with some of the assumptions that go into these ideas. Sometimes you don't like the conclusions that they lead to. Sometimes you just don't like the person expressing the alternative belief. Nevertheless, if your ideas are wrong, it's in your best interest to learn alternative concepts.

Be willing to consider alternatives that initially seem unpalatable. Suppose you believe that God created life and you learn that biologists have been able to put together a combination of the raw elements that existed on the young Earth along with energy, as from the young Sun, and create simple life.[4] So what do you do? Reject the scientific result as a fluke or just assume that it is another example of God's action? Or do you consider that there might be an alternative mechanism for the initial formation of life than the one you believe? Do you compare the two mechanisms, divine intervention and random evolution, on their merits? Do you accept that, by the principle of Occam's razor, the natural process for creating life requires fewer unproven assumptions and therefore warrants a change in your belief system?

---

[4] This hasn't happened yet, but the process has created some of the precursor molecules that go into making the more complex molecules that comprise living things.

8

Conflicts and Dangers

THE PROBLEMS THAT
MISCONCEPTIONS CREATE

The Rejection of Scientific Reality

A huge number of people are choosing to accept beliefs that are inconsistent with those put forward by science, or that presently lack scientific support. Among these beliefs are the presence of aliens from other worlds on Earth, the possibility of travel backward in time,[1] astrology,

---

[1] Based on Einstein's special relativity, it *is* possible to travel forward in time. As introduced in chapter 4, this possibility occurs because as things travel faster, their clocks tick more slowly as observed by the outside universe. In other words, if you went out in a spacecraft moving at close to the speed of light and then returned to Earth, people here would have aged much more than you. If you had traveled long enough and fast enough, you could come back thousands of years or more later having aged only a few years yourself.

angels, ghosts, demons, creationism, telepathy, extrasensory perception (ESP), magic, channeling and other methods of communicating with the dead, and homeopathic medicine, to name but a few. Accepting any of these alternative realities requires the believer to accept incorrect or unproven beliefs about the natural world.

"But what if I choose to live with my comfortable, if misconceived, beliefs about the natural world? I've done so for years and they haven't hurt me yet."

Once people decide to accept any nonscientific belief, they immediately begin adjusting the way they view the world so as to make it consistent with this new belief. I will work through the implications of two nonscientific beliefs, the presence of aliens on Earth and the use of homeopathic medicine, to show how any new incorrect belief can, in fact, hurt us, even without our knowing it.

Aliens in Our Midst

I enjoyed the movies *Men in Black* and *Independence Day*: nice escapism that speculates about different aspects of alien life visiting Earth today, a premise that I categorically reject, by the way. Let's consider what is involved when you actually accept that extraterrestrial aliens are here. In the first place, you must believe that they came from somewhere else, crossing space and arriving undetected by our current technology. The possibility that they are in the solar system undetected is plausible, since we are not actively monitoring all of it.

Since technology is a continually evolving field, I won't argue the possibilities or implications of interstellar travel other than to reiterate one science-related issue: there is no scientific evidence whatever that faster-than-light travel is possible, despite its occurrence in numerous science fiction stories. Indeed, there is very strong scientific evidence that the speed of light is truly the limit to how fast matter can travel. Therefore, it would take many years, more likely centuries or even millennia, for any aliens to reach the Earth. This implies that they almost certainly would come in gigantic spaceships carrying redundant quantities of everything they would need to survive such a long passage. The movie *Independence Day* had a good representation of such a ship.

If you believe aliens have made the journey, you have almost certainly asked yourself *why*. The usual explanations are that they have come either for positive reasons (from our point of view), such as sharing their knowledge with us; for neutral reasons, such as documenting life forms on Earth for their databases; or for negative reasons, such as taking raw materials from Earth or taking us as slaves, or worse.

One can rationalize why aliens have not heralded their presence to all humankind in either of the first two cases. Perhaps, for example, they don't think we are ready to know about them or they don't want to affect our cultural development.

If they are here to dominate us, then reasonable strategy dictates that they should either destroy us outright or use the element of surprise to overcome us before we could bring our ever-improving technology to bear against them. The argument that aggressive aliens are secretly studying us before they act doesn't hold any water because there are just too many chances for something to go wrong, enabling us to detect them. If they were here for domination, we would already know it by the destruction they caused. Since this hasn't happened, reason dictates that evil aliens aren't here . . . yet.

So if you believe aliens *are* here, you have probably decided that they are either just spying on us to learn about us or are waiting benevolently in the wings until we are ready to learn from them. Or you might have a completely different view about how evil aliens might be operating than I do. In any case, accepting the presence of aliens causes your view of the world to be different from mine. Mine does not contain aliens, and therefore I don't have to spend any intellectual energy thinking about them or their consequences.

If you also assume that aliens don't just stay in hidden landing craft (UFOs) but walk among us, then you must always be wondering on some level whether every stranger you meet is actually an alien. As I mentioned above, the chance of their being discovered is nontrivial, considering everything that could go wrong; since none of them have been paraded around, this leads to the question of government cover-up of alien bodies, craft, and artifacts. This conviction among many who believe in UFOs even has a focus, namely the restricted Area 51 in Groom Lake, Nevada.

Believing that there are aliens in our midst creates some degree of paranoia that affects how you interact with everyone. To blend in, aliens must either look strikingly like us or be able to disguise themselves fantastically well. If you believe they look like us, then this belief bears on your ideas about how humans developed. The theory of evolution clearly indicates that our present features evolved in response to our environment, and even so, we might very easily have evolved to look completely different than we do. For example, had circumstances been slightly otherwise, we could easily have four legs and four arms, with three fingers on each hand and eyes all around our head. The possibilities are virtually endless—even slightly different environments on the younger Earth would have led to profoundly different "humans." So if you believe aliens look even remotely like us, you must either assume that evolution occurs according to a much narrower set of guidelines than biologists believe it does, allowing creatures evolving even on different worlds to look similar, or conclude that evolution didn't happen and that all life throughout the galaxy was created by God "in His image."

Considering all the alleged alien contacts that occur here each year, it may well be that more than one alien race has arrived from different worlds. If so, are they mutual friends or foes? If they are enemies, then you also have to worry about them tearing up our solar system as they fight each other for dominance. See how complicated your worldview can become when you follow out the implications of a nonscientific belief?

WALK A MILE IN ANOTHER'S SHOES

As we each develop our own thought processes, we grow less and less able to appreciate other perspectives on deeply held or intellectually challenging beliefs. It is hard to step out of ourselves and look at the world from different points of view or from different belief systems. Therefore, we often don't know that our ideas are at odds with those of other people.

Sometimes, as when we assume that aliens are here on Earth, the effects of those beliefs are mostly behavioral. That is bad enough, because behaviors based on invalid assumptions are often not in our best interests. They might keep us associating with people we wouldn't otherwise, or keep us away from people or activities that would make our lives richer. Sometimes, as with belief in the validity of astrology, the effects are both behavioral and financial—causing people not only to make different decisions than they would otherwise but also to waste money. This is arguably worse than just missing opportunities to meet interesting people and have interesting experiences, but it is not nearly as bad as the effects of believing in such concepts as homeopathic medicine and faith healing. In such cases, people not only behave differently and waste money but also endanger their lives and the lives of people they love, especially children who may not have any choice in the treatment of their illnesses and injuries.

THE DOCTOR IS OUT

Homeopathic medicine asserts that various plants, minerals, and animal-derived compounds taken in very small doses will stimulate a sick person's natural defenses. The word "small" here is crucial. The concentrations of the substances are so small that often they contain only one allegedly medicinal molecule in an entire dose. There have been hundreds of studies, some scientific, some allegedly (but not actually) scientific, others clearly nonscientific, on the efficacy of homeopathic remedies. Aside from the blatantly nonscientific ones, some studies show positive health effects while others show no health effects at all from taking these remedies. Careful examination of the allegedly scientific studies reveals flaws in the ways they were conducted, so their results should not be accepted as scientifically valid.

It is the consensus among mainstream medical practitioners that any positive results from homeopathy are due to the placebo effect, the basic principle of which is that if you believe something is going to happen to you, your state of mind can influence the response of your body. This is, of course, precisely what happens when you take a homeopathic medicine: you believe, hope, or expect that you are taking something beneficial, and this belief starts physical activities within your body that help you heal—sometimes—or enables you to ignore pain or discomfort caused by whatever ails you. Similarly, some researchers studying the placebo effect suspect that the process of taking an inert placebo (thinking that it is actually a medicine) can help some people stimulate their bodies to react more to an illness or injury than they would otherwise.

The problems with homeopathy are, first, that belief in the healing power of a substance can help only some people get better, or at least feel better. It is much less effective than healing aided by medicines that have been shown scientifically to be effective. Therefore, when taking a homeopathic remedy, you have much less chance of healing as well or as quickly as you would by taking an established medicine. Second, homeopathic remedies are expensive; you will get better value for that money by spending it on standard medical care. Third, belief in homeopathy creates doubts about the value and validity of mainstream medicine. This latter point is very important. Since mainstream medicine has been proven significantly more effective than the placebo effect of homeopathy, reliance solely on homeopathic medicine can cause some people to put themselves at greater risk than if they followed strictly standard medical procedures. Much worse than choosing less effective remedies for themselves is withholding effective medical treatment from children or other people incapable of making such decisions for themselves.

Antiscience

Choosing to believe something inconsistent with current scientific consensus has the effect of drawing people away from the concepts that science teaches. Some people go even further and develop prejudices against science. This occurs for a variety of reasons, including:

- misunderstanding fundamentals about science and therefore drawing incorrect conclusions about what science says and does not say;
- refusing to change a personal belief inconsistent with science;

- developing a belief that there is more to physical reality than science can ever reveal;
- developing a belief that the scientific community is a monolithic organization out to ruthlessly crush anyone who opposes it, and then choosing to support the opposition;
- developing a belief that as a result of science, society is dehumanized; and, related to that,
- blaming science for many of the major problems that beset society, such as pollution, nuclear bombs, nuclear accidents, and nuclear waste.

IT'S MY RIGHT TO BE WRONG

People with nonscientific belief systems often become antagonistic to what science has to say and won't let correct information come between them and their ideas. I once participated in a public debate with a person who believed in the Steady State theory of the universe. He rejected the Big Bang theory for several reasons. First, he could not conceive of how the universe could have come into existence in the first place. This is a valid concern, since science does not yet have an explanation. Allowing that the observations of the universe expanding are correct, he concluded that it had existed forever and that therefore the Steady State theory explained how it could look the same for all time.

He also found it unacceptable that the Big Bang theory has had to be revised many times over the years, consistent with the basic tenet that science is an evolutionary process of theory, experiment, correction, and further experiment. To his way of thinking, the fact that each improved version of the Big Bang still had flaws meant that the entire construct should be rejected. Whereas scientists are perfectly happy to work with a promising model in order to make it consistent with reality, many nonscientists need to believe that any scientific theory they are told is entirely correct or entirely wrong.

His strongest argument against the Big Bang theory was aimed at the singularity at the beginning of time. He perceived the word "singularity" to mean a place where matter becomes infinitely dense. He got to this belief by using the mathematical principle that if you divide anything by zero (in this case, dividing the mass of the universe by zero diameter or size), you get infinity (in this case, infinite density). By the same reasoning, he also argued that black holes cannot exist because they are created when the mass concentrations become infinitely dense. His concept of a singularity is very common and, unfortunately, wrong.

To scientists, when mathematical equations "blow up" because they have zero as a divisor, it means that they fail to reflect reality accurately. Singularities, denoted by these equations, are places where science does not yet have an explanation for the nature of matter, energy, space, or time. Working with Einstein's equations of general relativity, astrophysicists have been able to work backward in time to a tenth of a millionth of a billionth of a billionth of a billionth of a billionth (10^{-43}) of a second after the universe began, before the equations fail due to their singularities. We don't yet have equations to take us from that early time back to the very beginning. Developing these, which will explain how the universe began, is one of science's "holy grails."

The public debate in which I participated lasted for nearly an hour and a half. During that time, I explained many science-related things, such as this concept of singularities. Judging from his responses and mannerisms, my opponent refused to accept any of these points[2]— they were too threatening to his personal cosmology.

PERSONAL SCIENCE

Some people blame scientists for many of the evils of the world (real and perceived), such as weapons of mass destruction, pollution, and dehumanization due to advanced technology. I believe that this blame is misdirected because it is based on the misconception that scientists know beforehand the implications to society of what they are going to discover. This is rarely true. One counterexample is the development

---

[2] If you have *absolutely nothing* better to do, you can order a videotape of the debate from Public Affairs, University of Maine, Orono, ME 04469 (207-581-3743) or www.umaine.edu/publicaffairs/.

of the atomic bomb, the Manhattan Project, during World War II. The process included both new engineering developments and the discovery of new scientific knowledge. The scientists working on it knew the potential destructive power of an atom bomb. But they had a war to win, and the faster they did it, they felt, the more lives would be saved.

Much of the science relevant to the creation of the first atom bomb was discovered long before its implications for weaponry were understood. Einstein, for example, had no idea in 1905 that his discovery of the relationship between energy and matter ($E = mc^2$) by splitting or fusing atoms could lead to the creation of bombs. Indeed, when he made his discovery, the structure of the atomic nucleus, necessary to understand how an atomic bomb would work, was not yet known. That only came into focus in 1920, when British physicist Lord Ernest Rutherford predicted the existence of electrically neutral particles in the nucleus. These "neutrons," discovered in 1932 by British physicist James Chadwick, provided a crucial piece of the nuclear puzzle needed to understand many aspects of nature, as well as to develop nuclear weapons, nuclear power, and nuclear medicine, among other applications. However, these came later and, at least in the case of nuclear weapons, were authorized and organized by governments, not scientists.

Indeed, the responsibility for scientific discoveries that may cause great harm rests primarily with the people empowered to make products from them. These decisions cannot be left just in the hands of scientists and engineers, or even corporations, because there will always be people willing to devise even weapons of mass destruction for the right price or the wrong reason. When the reason for making hazardous products is arguably noble, as with the development of the first atomic weapons during World War II, politicians, in this case President Roosevelt, not scientists, authorize and arrange for funding to develop them. Nowadays, however, their reasons are often unacceptable to the international community, since decisions in one country can affect all life on Earth. When any person, company, or state constructs dangerous products, it forces the rest of the human race to react to protect itself.

In the nuclear weapons example, after World War II, the awesome power the bomb had unleashed caused many scientists to lobby against further use of such weapons. The nuclear physicist J. Robert Oppenheimer, director of the Los Alamos National Laboratory, where the first atomic bombs were developed during World War II, told President Harry Truman, "Mr. President, I have blood on my hands." In defense of science as a process, let me observe that many of the subsequent problems related to the peacetime use of nuclear energy stemmed from engineering difficulties rather than lack of scientific understanding.

Engineering errors are a serious problem in every industry. Many of us are affected and most of us are angered when cars, for example, are recalled because of potentially life-threatening engineering defects. However, the only way to avoid such problems completely is to live in a world without technology or advanced medicine. I hope I have shown that the cost of such an alternative is unacceptable. Besides the medical ills that would befall most of us, we would all have far less opportunity to sustain or improve the quality of our lives. A positive perspective on our science and engineering–intensive society shows that products are continually being refined, making them less dangerous.

Instead of rejecting science, we can each choose which of its products to accept by buying them or not. If we do buy them, we are obligated to use them responsibly. As I write this book, the United States is debating the safety of using cellular phones in cars. There are advantages and disadvantages, but if you are hit by a car whose driver is distracted by his or her cell phone, then this particular product of high-tech engineering has a very personal impact on your life.

Scientists and the Scientific Process

The issue of whether a science-based view of the world is valid or not is complicated by widespread misunderstandings about what scientists *do*. Many people have developed incorrect beliefs about how we acquire scientific knowledge, why we scientists believe that knowledge, and how much everyone else is justified in believing it. It is interesting to note that some people give scientists more credit for understanding the natural world than they deserve, while others give scientists too little credit.

Too Much Credit

Many people believe that once a scientific discovery is made, that knowledge is somehow sacrosanct. This is rarely, if ever, true. Science consists of making observations or doing experiments, developing scientific theories to explain the results, testing the theories with other observations and experiments, and refining or replacing the theories as necessary.

There is nothing in the "scientific process" that allows for a scientific theory ever to be proven. If a theory makes a prediction shown to be consistent with observations or experiments, then out comes the champagne, but not because the theory has been proven—just because the theory is consistent with reality. There is never any guarantee that the next observation or the next experiment—with slightly different settings—will give similarly consistent results. In other words, *scientific theories can never be proven correct.*

Observations or experiments can support theories. These same observations or experiments can also disprove theories by showing them to be inconsistent with reality. When that occurs, and it frequently does, scientists must decide whether the theory should be revised or replaced. Since scientists are human, we don't like to let go of our treasured beliefs, and therefore we tend to make revisions until truly unassailable inconsistencies arise between the data and the treasured theory.

Even then, some theories persist until their advocates die, quite literally. As I indicated earlier, some people still advocate the Steady State model of the universe despite the fact that this theory makes predictions that are inconsistent with observations.

Too Little Credit

Undoubtedly, you'll have noticed that I've liberally used the word "theory" throughout the preceding chapters. In everyday usage, "theory" means an unproven idea, explanation, or method of doing something.

Normal usage of the word often creates the incorrect belief that scientific theories are just pie-in-the-sky ideas that may or may not be correct. Unlike scientific laws, scientific theories are perceived to lack concreteness. For example, "Newton's law of gravitation" sounds much more authoritative than "Einstein's theory of gravitation," even though Einstein's theory is much more powerful and accurate than Newton's law.

In science, the word "theory" means something completely different. A scientific theory is typically a group of ideas used to explain some physical phenomenon. A scientific theory can be expressed as one or more equations that can then be tested and modified, if necessary. Darwin's theory of evolution, Einstein's theories of relativity, and Dalton's atomic theory of matter all make predictions that can be, and have been, tested, and they have all been modified as necessary to conform with observations and experimental results.

Open to All

Another crucial property of science is that there are rarely any secrets about scientific knowledge kept from the public.[3] Indeed, scientists love it when other people learn more about their subject. Because we know how interesting and satisfying it is to understand how the universe functions, most of us are pleased to share our insights and feelings about scientific discoveries. Furthermore, the predictions made by scientific theories can in principle be tested by anyone willing to put in the time and effort to learn the theoretical background material and the methods of experimenting or observing. (In practice, many experiments are very expensive to do, so interested parties would need to get research grants just like the rest of us.)

Science is therefore an inclusive study in which everyone is invited to take part. Capable people, even without advanced college degrees,

---

[3] The exceptions to this rule are military and industrial secrets, but then these are not shared with outside scientists either. Secret projects mostly involve engineering information rather than pure science. Anyone suitably inclined to learn a scientific subject that might be presently known only to classified military or industrial scientists can still do so under different auspices, such as working at a university.

can do powerful and interesting things. For example, if you are an American citizen and you wanted to observe an object in space, you could apply for time at one of the national astronomical observatories. Your proposal would be evaluated on its scientific merits along with those submitted by professional research astronomers. If the review committee deemed your observations sufficiently important, you would be assigned time on a telescope.

WHITE LAB COATS AND PENCIL HOLDERS

A final reality check about scientists has to do with their personalities. There is no one type of person in science, and people come to it for different reasons. In the United States, roughly 80 percent of astronomers are men and 95 percent of astronomers are white. Despite all the stereotyped cartoons and stories, I have never seen an astronomer wear a pocket pencil holder, and I know of none who wear white lab coats except when working in clean conditions that require special clothing.[4]

While in astronomy there most certainly are people you would classify as nerds, the vast majority are exceptionally interesting and well-rounded. Among my friends and acquaintances in the field are members of rock bands, black belts in karate, recreational and marathon runners, successful authors, dancers, hikers, bird watchers, gardeners, fishermen, gourmands, mountain climbers, spelunkers, sailors and powerboat enthusiasts, stock market players, and poets.

Being an astronomer certainly does require above-average skills in mathematics and physics, and these skills affect how we look at the world. This does not mean that we and other scientists are necessarily boring eggheads. In fact, my experience is that the more someone understands about how any aspect of the natural world truly works, the more interesting they are to talk to.

Let's now consider the question raised earlier about whether we should believe scientific theories such as the Big Bang just because

---

[4] Lab coats serve a functional importance in other fields, protecting people from chemicals or biological agents. The real issue is what the people are like when the lab coats come off.

most astronomers do. When you hear of a theory accepted by the scientific community to explain some aspect of nature, you can be sure that a number of observations and experiments consistent with the theory's predictions have taken place. The Big Bang theory of cosmology makes predictions that are all consistent with observations, and to date, *no other theory of cosmology is consistent with all existing observations.* This is why nearly all astronomers believe it, and that is a compelling reason for you to believe it, if you choose.

Mental Firewalls

Both in science and in everyday life, the process of changing our belief system is so hard, painful, and protracted that we often end up retaining invalid prior beliefs while accepting new, possibly contradictory, information. To do this, we build a mental firewall between the contradictory elements of the two sets of beliefs and only breach it to reconcile concepts when absolutely forced to do so. Indeed, from childhood on, we often have competing, incompatible beliefs that live side by side in our minds. For example, I know devoutly religious scientists who spend their professional careers developing scientific theories that are antithetical to the teachings of their faiths.

We humans are bundled contradictions. We are remarkably comfortable with conflicting beliefs. We draw conclusions using invalid reasoning and illogical, emotional gut feelings. We trust unreliable sources and act upon beliefs based on falsehoods that we once learned as gospel truth. We do things we know aren't good for us, whether it is speeding down a highway too fast or drinking too much. In such instances, our better judgment is overridden by our search for thrills or escape from painful reality. We rationalize many things, like pollution from motor vehicles, rather than correcting the problems—until the damage is so great that it can't be ignored. Often, we manage to survive and even prosper without a scientific appreciation of nature. But is it enough? Is our acceptance of incorrect information, from which we build our views of the world, the best we can expect from life? I think we can do better.

Epilogue

FALSE PERSONAL COSMOLOGIES

By the time most of us are in our teens, we have cobbled together a set of beliefs about how and why the universe began, how the Earth formed, why we are here as individuals, and other "big-ticket" issues. I call this a personal cosmology. Most personal cosmologies are different from the general cosmology under development by astrophysicists and other scientists. Personal cosmologies are shaped by each person's unique experiences growing up and, often, by religious training. For example, many people believe that the Earth formed as a direct result of the Big Bang explosion that created the universe, rather than some eight billion years later out of the debris of previous stars. Other beliefs include that God created the Earth in six days, that the Earth is only six thousand years old, and that a flood of biblical proportions covered the Earth. These beliefs are inconsistent with the cosmology based on astronomical and geological observations and associated scientific theories. Sometimes religion-based precepts remain with people throughout their lives, but nowadays we have so many other sources of information and beliefs that personal cosmologies frequently evolve or even change radically as we grow.

Every year some fundamentalist Christian students take my introductory astronomy course. Through in-depth interviews, I have learned that some of them take everything I say that contradicts their religious beliefs as just an alternative belief system, learn it for the exams, and promptly forget it afterward. Others have great difficulty dealing with an astronomer's version of cosmic events. Several students have talked with me about how conflicted they felt at the end of the course about what had previously been purely theological issues for them, such as the formation of the Earth and the evolution of life. Discovering that science has developed a vast system of theories that can explain many things without the need for God's intervention disturbed them profoundly.

The contrast between the logic inherent in scientific theories, along with their predictive power and internal consistency, and the unpredictable and inconsistent theological explanations of cosmological events can create great internal tension in people who are willing to consider both kinds of theories honestly. Many of the students I spoke to expressed great anger, sometimes toward me but usually toward their priests, ministers, or parents, who had forced them to construct personal cosmologies that they were coming to believe to be false. I have never followed up with these religious students to see if any of them ever replaced or significantly revised a personal cosmology because of what they learned in college. It would be interesting to find out.

Getting to There from Here

Personal cosmologies are rarely complete, but most of us use what we believe is true to generate answers when we are asked questions about the cosmos (or indeed about any subject) that we may have never considered before. For example, suppose I were to ask you where in the Milky Way galaxy our solar system is located. Common answers include in the center (incorrect but plausible, because we see the Milky Way extending all around the Earth); in a spiral arm (incorrect but plausible, for the same reason); and outside the Milky Way (incorrect but plausible, because we don't see any stars "near" us).[1] In fact, the solar system is located between two spiral arms.

Other examples of incorrect information based on false personal cosmologies are that all the stars are the same distance from the Earth; that all the stars in the galaxy are visible to the naked eye (for every star we can see, there are 30 million stars in the galaxy we can't see); that stars exist forever (they shine for between a few tens of millions of years to over a hundred billion years, depending on their mass); that stars don't evolve (they do—in fact, all the elements in your body except hydrogen, helium, and lithium were made inside stars that shed much of their contents back into space in supernova explosions); that star formation ceased billions of years ago (we observe it occurring today); that there are many fewer stars than actually exist (at least 10,000 billion billion stars exist in the universe); that there is only one galaxy in the universe; and that there are only a few galaxies in the universe.

Unlike religion, which demands your trust and promises you absolute truth in return, science encourages you to think critically for yourself and only promises to help sort out plausible ideas from implausible ones. It has shown, for example, that the Steady State universe model is not viable, while the Big Bang model is. Although the Big Bang creates a conceptual framework for understanding the evolution of the cosmos, there is absolutely no guarantee that all the presently accepted details are correct. Indeed, it is virtually certain that some are not. So what is the advantage of a scientific viewpoint about

---

[1] "Near" is a relative term, of course. When we look up at the night sky, the distances to stars seem very large compared to everything else with which we are familiar. The impression we get is often of great isolation of the Earth from the rest of the cosmos. Despite the stars' apparent remoteness, most people still underestimate their distances, usually giving estimates of millions or billions of miles or kilometers. In fact, the closest star, Proxima Centauri, is 25 trillion miles (40 trillion kilometers) from Earth.

the natural world, which can never be absolutely certain of anything, over one built on a foundation of religious faith? It is intellectual freedom—the right to question everything and everyone about their assertions and beliefs, to examine every aspect of the natural world, and to propose your own ideas and see them tested. Recognizing and overcoming our misconceptions is the first step in freeing our minds so that we can understand how our world works and use this knowledge to take control of our lives.

SELECTED BIBLIOGRAPHY

Armstrong, Thomas. *Multiple Intelligence in the Classroom*. Alexandria, Va.: Association for Supervision and Curriculum Development, 1994.

Arons, Arnold B. *A Guide to Introductory Physics Teaching*. New York: Wiley, 1990.

Bad Astronomy. www.badastronomy.com

Baron, Joan Boykoff and Robert J. Sternberg, eds. *Teaching Thinking Skills: Theory and Practice*. New York: W. H. Freeman, 1987.

Burnham, John C. *How Superstition Won and Science Lost: Popularizing Science and Health in the United States*. New Brunswick, N.J.: Rutgers University Press, 1987.

Cromer, Alan. *Uncommon Sense: The Heretical Nature of Science*. New York: Oxford University Press, 1993.

Damer, T. Edward. *Attacking Faulty Reasoning: A Practical Guide to Fallacy-Free Argument*. 4th ed. Belmont, Calif.: Wadsworth, 2000.

Gardner, Howard. *Multiple Intelligences: The Theory in Practice*. New York: Basic, 1993.

———. *The Unschooled Mind: How Children Think and How Schools Should Teach*. New York: Basic, 1991.

Gardner, Martin. *Science: Good, Bad and Bogus*. Buffalo, N.Y.: Prometheus, 1990.

Gentner, Dedre and Albert L. Stevens, eds. *Mental Models*. Hillsdale, N.J.: Lawrence Erlbaum Associates, 1983.

Giere, Ronald N. *Understanding Scientific Reasoning*. 4th ed. New York: Harcourt, 1998.

Gilovich, Thomas. *How We Know What Isn't So: The Fallibility of Human Reasoning in Everyday Life*. New York: Free Press, 1991.

Gleick, James. *Faster: The Acceleration of Just About Everything*. New York: Pantheon, 1999.

Gouguenheim, Lucienne, Derek McNally, and John R. Percy, eds. *New Trends in Astronomy Teaching: IAU Colloquium 162*. Cambridge: Cambridge University Press, 1998.

Harding, T. Swann. *The Joy of Ignorance*. New York: William Godwin, 1932.

Hawking, Stephen W. *A Brief History of Time: From the Big Bang to Black Holes*. New York: Bantam, 1988.

Helm, Hugh and Joseph Novak. *Proceedings of The International Seminar on Misconceptions in Science and Mathematics*, June 20–22. Ithaca, N.Y.: Cornell University Press, 1983.

Holton, Gerald. *Einstein, History and Other Passions: The Rebellion Against Science at the End of the Twentieth Century*. New York: Addison-Wesley, 1996.

Jastrow, Joseph, ed. *The Story of Human Error*. Freeport, N.Y.: Books for Libraries Press, 1936.

Lipps, Jere H. *Beyond Reason: Science in the Mass Media*. In J. W. Schopf, ed., *Evolution! Facts and Fallacies*. San Diego: Academic Press, 1999.

Mackay, Charles. *Extraordinary Popular Delusions and the Madness of Crowds*. 1841; reprint, New York: Farrar, Straus & Giroux, 1932.

Nelkin, Dorothy. *Selling Science: How the Press Covers Science and Technology*. Rev. ed. New York: W. H. Freeman, 1995.

Nickell, Joe. *Entities: Angels, Spirits, Demons, and Other Alien Beings*. Buffalo, N.Y.: Prometheus, 1995.

Nickerson, Raymond S. *Reflections on Reasoning.* Hillsdale, N.J.: Lawrence Erlbaum Associates, 1986.

Nickerson, Raymond S., David N. Perkins, and Edward E. Smith. *The Teaching of Thinking.* Hillsdale, N.J.: Lawrence Erlbaum Associates, 1985.

Novak, Joseph. *Proceedings of the Second International Seminar on Misconceptions in Science and Mathematics,* July 26–29. Ithaca: Cornell University Press, 1987.

——. http://www2.ucsc.edu/mlrg/mlrghome.html

——. http://www2.ucsc.edu/mlrg/proc1abstracts.html

——. http://www2.ucsc.edu/mlrg/proc2abstracts.html

——. http://www2.ucsc.edu/mlrg/proc3abstracts.html

——. http://www2.ucsc.edu/mlrg/proc4abstracts.html

Odenwald, Sten. *The Astronomy Café: 365 Questions and Answers from "Ask the Astronomer."* New York: W. H. Freeman, 1998.

Paulos, John Allen. *Innumeracy: Mathematical Illiteracy and Its Consequences.* New York: Vintage, 1990.

Pirsig, Robert. *Zen and the Art of Motorcycle Maintenance.* New York: Bantam, 1984.

Randi, James. *Flim-Flam! Psychics, ESP, Unicorns, and Other Delusions.* Buffalo, N.Y.: Prometheus, 1988.

Rothman, Milton A. *The Science Gap: Dispelling the Myths and Understanding the Reality of Science.* Buffalo, N.Y.: Prometheus, 1992.

Rutherford, F. James and Andrew Ahlgren. *Science for All Americans: Education for a Changing Future.* New York: Oxford University Press, 1990.

Sadler, Philip Michael. *The Initial Knowledge State of High School Astronomy Students.* Ph.D. diss., Graduate School of Education, Harvard University, 1992.

Sagan, Carl. *The Demon-Haunted World: Science as a Candle in the Dark.* New York: Random House, 1995.

Sagan, Carl and Thornton Page, eds. *UFOs—A Scientific Debate.* Ithaca, N.Y.: Cornell University Press, 1972.

Schick, Theodore, Jr. and Lewis Vaughan. *How to Think About Weird Things: Critical Thinking for a New Age.* 2nd ed. Mountain View, Calif.: Mayfield, 1999.

Schnabel, Jim. *Round in Circles: Physicists, Poltergeists, Pranksters, and the Secret History of the Cropwatchers.* Buffalo, N.Y.: Prometheus, 1994.

Scott, P. H., H. M. Asoko, and R. H. Driver. *Teaching for Conceptual Change: A Review of Strategies.* In R. Duit, F. Goldberg, and H. Niederer, eds.,

Research in Physics Learning: Theoretical Issues and Empirical Studies. Proceedings of an International Workshop. IPN 131, 310–329. Kiel, Germany: Institute for Science Education, March 1991.

Seeds, Michael A. *Foundations of Astronomy: 1999 Edition.* Belmont, Calif.: Wadsworth, 1999.

Shermer, Michael. *Why People Believe Weird Things: Pseudoscience, Superstition, and Other Confusions of Our Time.* New York: W. H. Freeman, 1997.

Stepans, Joseph. *Targeting Students' Science Misconceptions: Physical Science Activities Using the Conceptual Change Model.* Riverview, Fla.: Idea Factory, 1994.

Thorne, Kip S. *Black Holes & Time Warps: Einstein's Outrageous Legacy.* New York: Norton, 1994.

White, Richard T. *Learning Science.* Oxford: Basil Blackwell, 1988. (advanced book on education theory)

Wiggam, Albert Edward. *Sorry But You're Wrong About That.* Indianapolis, Ind.: Bobbs-Merrill, 1931.

Wolpert, Lewis. *The Unnatural Nature of Science.* Cambridge: Harvard University Press, 1992.

Ziman, John. *The Force of Knowledge: The Scientific Dimension of Society.* Cambridge: Cambridge University Press, 1976.

——. *Reliable Knowledge: An Exploration of the Grounds for Belief in Science.* Cambridge: Cambridge University Press, 1978.

——. *Teaching and Learning About Science and Society.* Cambridge: Cambridge University Press, 1980.

INDEX

authority, 199, 201; questioning traditional beliefs, 202

dark matter, 98
day, length of: on Earth, 14 (figure), 17, 19, 108; on Mercury, 30
Deimos, 102, 189 (figure)
density misconceptions, 22, 117
dinosaurs, 50–51
disease, 129–134, 136–137
DiSessa, Andrea, 103, 106
Doppler shift, 161

Earth: as center of the universe, 47, 90; length of day, 14 (figure), 17, 19, 108; length of year, 5–6, 7; lifespan of, 108–109; population statistics, 69; rotation of, and tides, 44, 45 (figure); seasons on, 12–19, 103; shape of orbit, 13; slowing of rotation, 44; temperature of, 29, 30; tides, 38–44, 42 (figure); tilt of axis of rotation, 13–17, 14–15 (figures), 19; and uniqueness assumptions, 107
Earthshine, 38
eclipses, lunar, 34–38
Eddington, Sir Arthur, 22
education. See teachers; teaching techniques; textbooks
Einstein, Albert, 161–162, 223
Einstein's equation $E = mc^2$, 24, 108, 223
electromagnetic spectrum: and astrology, 67; color of Sun's light, x, 46, 86–89, 87 (figure), 109
The Empire Strikes Back, 10–11, 52

energy misconceptions: conservation of energy, 23–24; source of Sun's energy, 5, 6, 19–24, 108–109
epicycles, 156, 157 (figure)
epidemics, 136–137
Europa, 205

falling objects, 83, 116–117, 201; *see also* trajectories
falling stars, 46, 64
faster-than-light travel, 51–52, 84–85
fission, 24
The Flintstones, 50–51
forces, fundamental, 71–72
friction, 116, 117
"frozen" gases, 72
fusion: cold, 201; and novas, 64; and solar neutrinos, 200; as source of Sun's energy, 24, 108–109; in stars, 120

galaxies: defined, 64; confused with solar system, 61; number of, 47; number of stars in spiral arms vs. "empty" space, 90–92, 91 (figure); red shift, 161; shape of, 47
Galileo Galilei, 158–159, 201
general relativity, 60, 161–162
global warming, 143–144
God, belief in, 214, 218, 229–232
Gold, Thomas, 63
gravity, 47, 49–50; antigravity devices, 51–52; and astrology, 66; and expanding universe, 161–162; falling objects, 83, 116–117, 201; limits of Newton's theory, 60; Newton's laws, 159–160; and trajectories, 50, 100, 116–117